16th International Conference: Mechatronic Systems and Materials (MSM 2021)

AA003737

IOP Conference Series: Materials Science and Engineering
Volume 1239

Vilnius, Lithuania
30 June - 2 July 2021

Editors:

Vytautas Bucinskas
Dainius Udris

ISBN: 978-1-7138-5440-1
ISSN: 1757-8981

Printed from e-media with permission by:

Curran Associates, Inc.
57 Morehouse Lane
Red Hook, NY 12571

CURRAN ASSOCIATES INC.
proceedings.com

Some format issues inherent in the e-media version may also appear in this print version.

This work is licensed under a Creative Commons Attribution 3.0 International Licence.
Licence details: http://creativecommons.org/licenses/by/3.0/.

No changes have been made to the content of these proceedings. There may be changes to pagination and minor adjustments for aesthetics.

Printed with permission by Curran Associates, Inc. (2022)

For permission requests, please contact the Institute of Physics at the address below.

Institute of Physics
Dirac House, Temple Back
Bristol BS1 6BE UK

Phone: 44 1 17 929 7481
Fax: 44 1 17 920 0979

techtracking@iop.org

Additional copies of this publication are available from:

Curran Associates, Inc.
57 Morehouse Lane
Red Hook, NY 12571 USA
Phone: 845-758-0400
Fax: 845-758-2633
Email: curran@proceedings.com
Web: www.proceedings.com

TABLE OF CONTENTS

IOP Conf. Series: Materials Science and Engineering 1239 (2022) 011001 doi:10.1088/1757-899X/1239/1/011001

The 16th International Conference: "Mechatronic Systems and Materials" (MSM 2021), 01-02 July 2021

The 16th International Conference "Mechatronic Systems and Materials" (MSM 2021) is being organized by:

- Mechanical Faculty of Vilnius Gediminas Technical University
- Kaunas University of Technology
- Lithuanian Academy of Sciences
- Opole University of Technology
- Bialystok Technical University
- IFToMM National Committee of Lithuania

Aim & Scope

Providing opportunity for the scientists to share information and facilitate co-operation in Mechatronics, New Materials, and Engineering Technologies and dissemination of current research results in this multi-disciplinary field.

TOPICS

Mechatronic Systems:

- Robotics: Industrial, Microrobotics, Mobile Robots
- Applications of Artificial Intelligence
- Sensors and Actuators in Mechatronics
- Control of Mechatronics Systems
- Analysis of Vibration
- Optimization, Optimal Design
- Integrated Diagnostics
- Failure Analysis
- Tribology in Mechatronics Systems
- Analysis of Signals

Materials:

- Multifunctional and Smart Materials
- Metallic Alloys
- Piezoelectric Materials
- Nanomaterials
- Ceramics and Glasses
- Biomaterials and technology

Engineering Technologies:

- Advanced and digital manufacturing
- Systems engineering
- Micro and nano technologies
- Materials coating and joining technologies
- Modeling and optimization of processes

Education:

- New Trends and Curricula for BSc and MSc in Two Tier Higher Education in the Fields of Mechatronic
- Systems and Materials Science.

Content from this work may be used under the terms of the Creative Commons Attribution 3.0 licence. Any further distribution of this work must maintain attribution to the author(s) and the title of the work, journal citation and DOI.
Published under licence by IOP Publishing Ltd

IOP Conf. Series: Materials Science and Engineering 1239 (2022) 011001 doi:10.1088/1757-899X/1239/1/011001

Scientific Editorial Committee

Chairman: **Vytautas BUČINSKAS**, Lithuania. Head of Department of Mectaronics, Robotics and Digital Manufacturing; Faculty of Mechanics; Vilnius Gediminas Technical University
Vice Chairman: **Dainius UDRIS**, Lithuania. Vice Dean of Faculty of Electronics; Vilnius Gediminas Technical University

Members

Algirdas Vaclovas VALIULIS, Lithuania
João CALADO, Portugal
Artūras SERACKIS, Lithuania
Valentin ANTONOVIČ, Lithuania
Justinas GARGASAS, Lithuania
Algimantas BUBULIS, Lithuania
Sebastian DENDORFER, Germany
Lars KRENKEL, Germany
Alessandro RUGGIERO, Italy
James SHIPPEN, United Kingdom
Rymantas Tadas TOLOČKA, Lithuania
Vadim MOKŠIN, Lithuania
Konrad LEWCZUK, Poland
Egidijus DRAGAČIUS, Lithuania
Lech DZIENIS, Poland
Konrad J. WALUS, Poland
Florin NEMTANU, Romania
Tadeusz LAGODA, Poland
Inga MORKVĖNAITĖ-VILKONČIENĖ, Lithuania
Olegas ČERNAŠĖJUS, Lithuania
Algirdas MAKNICKAS, Lithuania
Jelena ŠKAMAT, Lithuania
Sergejus BORODINAS, Lithuania
Yuri DEKHTYAR, Latvia
Ching-Chi HSU, Taiwan
Luis MOCHIZUKI, Brasil
Andrius DZEDZICKIS, Lithuania
Małgorzata SYCZEWSKA, Poland
Bart VERKERKE, The Netherlands
Adrian BURLACU, Romania
Miroslaw KARCZEWSKI, Poland
Pietro GIANNATTASIO, Italy
Cevdet GOLOGLU, Turkey
Abdallah ALZOUBI, Jordan
Hikmet KARAKOC, Turkey
Roland PAWLICZEK, Poland

Organizing Committee

Chairman: Ernestas ŠUTINYS, Lithuania
Vice Chairmans: Sigitas PETKEVIČIUS, Lithuania; Vytautas TURLA, Lithuania
Members: Eglė PASTARNOKIENĖ, Lithuania; Irina GRINBERGIENĖ, Lithuania; Deividas MIZERAS, Lithuania; Ieva ŠVAGŽDYTĖ, Lithuania; Tadas LENKUTIS, Lithuania; Marius ŠUMANAS, Lithuania

IOP Conf. Series: Materials Science and Engineering 1239 (2022) 011002 doi:10.1088/1757-899X/1239/1/011002

Peer Review Statement

All papers published in this volume have been reviewed through processes administered by the Editors. Reviews were conducted by expert referees to the professional and scientific standards expected of a proceedings journal published by IOP Publishing.

- **Type of peer review:**
- **Conference submission management system:** Morressier
- **Number of submissions received:** 27
- **Number of submissions sent for review:** 27
- **Number of submissions accepted:** 18
- **Acceptance Rate (Submissions Accepted / Submissions Received × 100):** 66.7
- **Average number of reviews per paper:** 2
- **Total number of reviewers involved:** 15
- **Contact person for queries:**
  Name: Ernestas Sutinys
  Email: ernestas.sutinys@vilniustech.lt
  Affiliation:

Content from this work may be used under the terms of the Creative Commons Attribution 3.0 licence. Any further distribution of this work must maintain attribution to the author(s) and the title of the work, journal citation and DOI.
Published under licence by IOP Publishing Ltd

IOP Conf. Series: Materials Science and Engineering 1239 (2022) 012001 doi:10.1088/1757-899X/1239/1/012001

Advanced ICT solutions in a distance vocational education and training in the field of robotics

Zbigniew Pilat[1, 6 *[0000-0002-4553-1695]], Polyxeni Arapi[2], Nikos Pappas[2], Mirosław Omieljanowicz[3], Norbert Prokopiuk[4], Mart Tamre[5]

[1] ŁUKASIEWICZ - Industrial Research Institute of Automation and Measurements PIAP, Warsaw, Poland
[2] Technical University of Crete, Laboratory of Distributed Multimedia Information Systems and Applications (TUC/MUSIC), Chania, Greece
[3] Bialystok University of Technology, Faculty of Computer Science, Bialystok, Poland
[4] Warsaw University of Technology, Faculty of Power and Aeronautical Engineering, Warsaw, Poland
[5] Tallinn University of Technology, School of Engineering, Tallinn, Estonia

[6] E-mail: zbigniew.pilat@piap.lukasiewicz.gov.pl

Abstract. Automation and robotization (A&R) is often the main driving force behind the development of advanced manufacturing, including the transformation of the company towards the Industry 4.0 model. Effective implementation and effective use of A&R solutions is possible only if the company has good prepared and educated employees. It applies to all levels, from the direct operation of machines and processes, through mid-level technical staff, to the management level. Nowadays, education of young staff, as well as training of those currently employed, requires the implementation of new solutions and methods of Vocational Education and Training (VET). They should ensure not only a high professional level, but also meet the challenges facing VET today: attractiveness of education for students, especially young ones, universal accessibility regardless of the place of residence, internationalization. The paper, presenting results of the MILAN project, shows that addressing those problems can largely be facilitated by the use of modern, advanced ICT solutions.

1. Introduction

Access to effective and attractive educational activities and vocational training in the field of robotics faces many barriers. The problem concerns both students at technical secondary schools, students of professional development courses, students at universities and employees of industrial companies. One of the limitations is access to practical exercises with the robot, where demonstration of its operation is possible. Currently, an education unit or industrial company that wants to train its employees must send them to a training center equipped with robotic training stations or organize such a training cell at its own. It is costly, time-consuming and limits access to training for many potential recipients. The pandemic that has prevailed for over a year has significantly aggravated those problems. Laboratories are closed; travel opportunities for both participants and training staff are limited. For this reason, there is an increasing demand for all forms of distance learning, including vocational training.

One of the goals of the MILAN (Erasmus +) project is to develop and pre-test such a new approach that uses distance learning methods supported by advanced ICT solutions [1]. Now, after the outbreak

Content from this work may be used under the terms of the Creative Commons Attribution 3.0 licence. Any further distribution of this work must maintain attribution to the author(s) and the title of the work, journal citation and DOI.
Published under licence by IOP Publishing Ltd

of the pandemic, this part of the project seems to be gaining in importance. The article presents the preliminary results of the project, including an online education and training platform that integrates a series of e-learning courses on Robotics and a multimedia library. Beyond traditional multimedia objects (such as photos, videos etc.), the library includes 3D objects and simulations that the student can experience in AR or VR. A wide range of devices are supported, from computer stations equipped with AR / VR systems and headsets to popular mobile devices, e.g. smartphones.

2. About the MILAN project

MILAN project [2] is directly addressed towards the Erasmus+ call 2018 Horizontal priority "Open and innovative practices in a digital era". The project is run to elaborate high-quality, open-access and innovative training materials and tools in the field of Robotics and Automation, involving VR and AR technologies. All MILAN training materials are going to be uploaded to an interactive e-learning platform, which will enable a new quality of training, oriented towards the personal needs of the users. It is targeted to the students tending to develop in the field of Advanced Manufacturing. The MILAN training system is supposed to contain online e-learning courses, lessons for mobile devices, interactive exercises and movies available either from PCs, tablets or smart-phones. MILAN's outcomes are also going to involve modern ICT solutions, including Virtual and Augmented Reality.

The main target groups of the project are:

- people employed at SMEs: operators of advanced machines (robots, automats, assembling cells/machines) and middle-level technical staff,
- teachers/trainers/consultants in the field of Robotics and Automation,
- students of vocational schools.

3. MILAN online multifunctional education and training platform

MILAN will offer an online education and training platform that integrates a series of e-learning courses on Robotics and a rich multimedia library. It is built on Coursevo, a multilingual multimedia educational platform for managing courses and supporting learning processes and learning communities through the Web, developed by the Technical University of Crete /Laboratory of Distributed Multimedia Information Systems and Applications [3].

Coursevo fosters distance learning by enabling communication between tutors/trainers and students, cooperation among students and access to coursework information and learning resources. It can combine traditional classroom-based lessons and practical sessions, with self-study and eLearning. This, so called, "hybrid" or "blended" approach provides a significant learning opportunity as it combines the immediacy of communication between educators and learners with the irreplaceable practical training in laboratories and the convenience, flexibility and self-regulation of personal study without time/space constraints.

The platform also offers features and tools for communication and collaboration and supports the creation of virtual classrooms and remote connection with laboratories, such as the ŁUKASIEWICZ – Industrial Research Institute for Automation and Measurements PIAP (access to the camera in KUKA robot's cell) and Technical University of Košice (modeling and virtual laboratory) in MILAN project.

The MILAN online training platform is accessible from: https://milan.coursevo.com/. Its graphical user interfaces are responsive, so it can be accessed and used both from desktop and mobile devices (Figure1). One can log in to the platform using his/her existing Facebook or Google account or sign up for an account in the platform. Then, (s)he can register in MILAN courses for free and follow course activities, communicate and collaborate with other users etc.

The training programme of the MILAN project is divided into four blocks: 1) Basics of Advanced Manufacturing, 2) Automation and Robotics in Advanced Manufacturing, 3) ICT in Advanced Manufacturing, and 4) Occupational safety and health. Under those blocks, a number of courses have been developed, divided into lessons (Figure 1, Figure 2). Each lesson is accompanied by a textbook. The lesson training content also includes multimedia content and experiences supported by AR and VR technologies (see next section), resources for further study, assignments to apply the theoretical knowledge gained in practice, and a self-assessment consisting of multiple choice questions to assess student's knowledge on the subject. Learning activities are supported by the platform services, including services and tools for communication and collaboration. The educators are able to activate the proper subset of course services depending on the course needs.

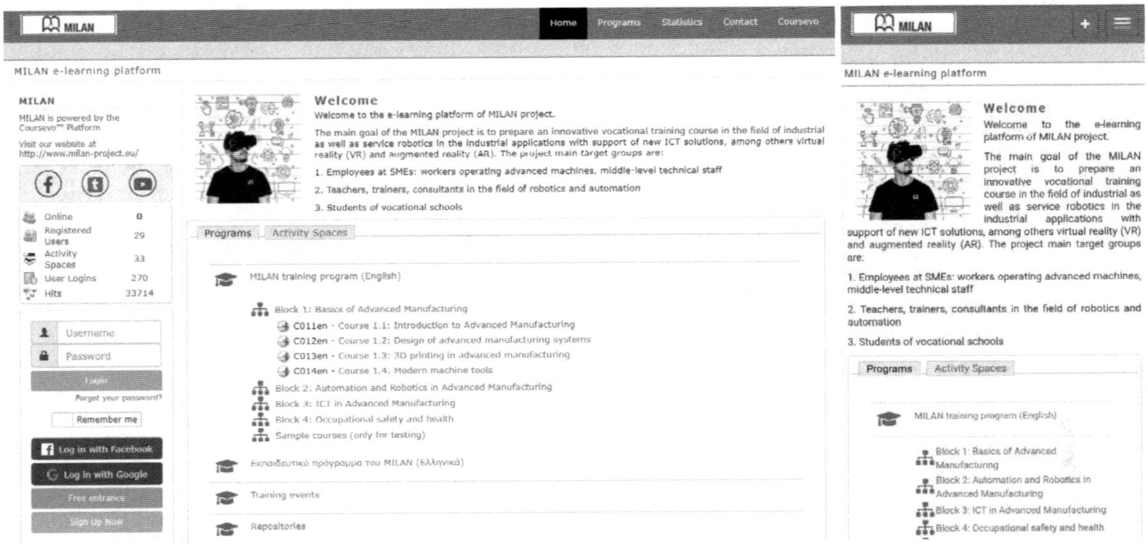

Figure 1. Main page of MILAN Training Platform (desktop and mobile interface).

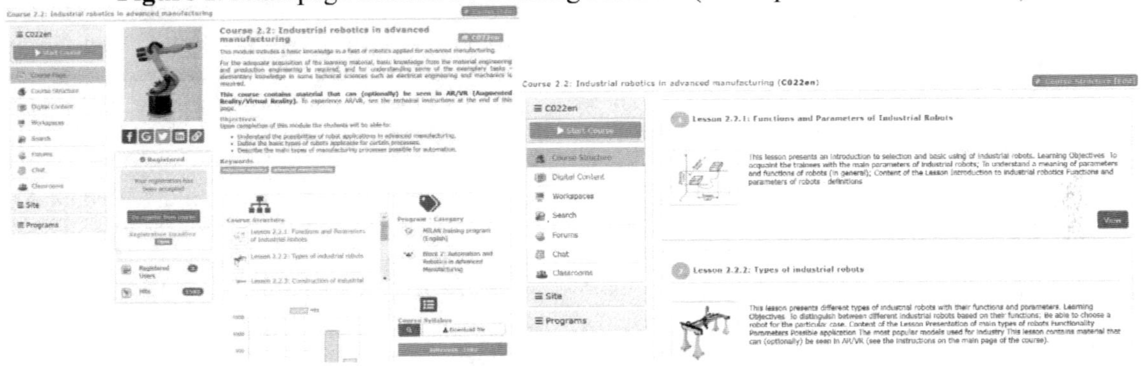

Figure 2. Course structure.

4. Support and use of multimedia

Learning activities using multiple media can be more effective than doing it through a single medium (such as text). The most effective multimedia provides learning experiences that mirror real-world experiences and let learners apply the content in various contexts.

This is aligned with MILAN approach, where multimedia is used as an important asset to support learning activities and practical training in the field of Robotics and Automation. Thus, the lessons offered in the MILAN Training System are not only based on text content, but they also use multimedia materials stored in a multimedia library. Beyond traditional multimedia objects (such as graphics, photos, videos, presentations etc.), the library includes 3D objects, simulations and applications that the student can experience in AR or VR.

Coursevo gives emphasis in the use of multimedia as a powerful learning means offering services for their efficient management and use in learning activities [3]. Among others, the platform allows integrating multimedia objects into any part of a course structure, also including web-based immersive AR/VR experiences, which can be delivered from the platform and the browser without needing to install any native applications (although this is also possible) (Figure 3, Figure 4). This not only applies to multimedia uploaded to the platform but also to external resources, which are also used in MILAN. A course author is able to attach external resources to learning activities, e.g. from YouTube, SlideShare, or any web site, that will be presented to the student through the platform, without leaving the course or open any new tabs or applications.

Figure 3. Integration of traditional multimedia into lessons.

5. Building and integrating AR and VR experiences into lessons

MILAN focuses on the use of web-based technologies and formats for building and providing AR and VR experiences through its training platform, allowing this way easy use and interoperability with a wide range of devices, from computer stations equipped with AR / VR systems and headsets to popular mobile devices, e.g. smartphones. Towards this end, special functionality has been developed in the platform and web-based APIs have been exploited for the implementation of AR/VR experiences on the Web, such as WebAR/VR and the newest WebXR Device API [4]. In MILAN, AR/VR experiences may include presentation of 3D objects or animated simulations (i.e. demonstrating robots parts or function and manufacturing processes), as well as applications supporting complex scenarios and user interactions useful in practical training (e.g. interactive simulations).

For the needs of MILAN project, a new functionality has been developed in Coursevo, which allows uploading 3D models and animated simulations in glb format and their presentation on the Web, in AR or VR. GLB describes 3D models/scenes in a standard format optimized for the Web. It is used in AR, VR, games, and web applications, and supports motion and animation and fast load times. GLB files are the binary version of the GL Transmission Format (glTF), an API-neutral runtime asset delivery format developed by Khronos Group [5]. glTF bridges the gap between 3D content creation tools and modern graphics applications by providing an efficient, extensible, interoperable format for the transmission and loading of 3D content. The glTF/glb format is supported by lots of software. Moreover, software that is widely used in Robotics, such as ABB RobotStudio and Solidworks, allow exporting simulations and models to glTF/glb format in their recent versions, so that they can be used in AR, VR and web experiences. There are also lots of libraries with 3D models in this format.

Such models and simulations in glb format can be uploaded by a course author to Coursevo and integrated into any part of a course structure, without needing to write any code for their presentation on the Web, in AR or VR. At the presentation time, the component developed does the background job to "wrap" the 3D object with the needed code to be presented on the web inside the course structure, and allow the user to interact with it (e.g. rotate it, play/stop if it is animated etc.), and view it in AR or VR (Figure 4). The implementation exploits Google's Scene Viewer [6] and Babylon.JS viewer [7] to support web-based AR/VR experiences across a wide range of devices.

IOP Conf. Series: Materials Science and Engineering 1239 (2022) 012001 doi:10.1088/1757-899X/1239/1/012001

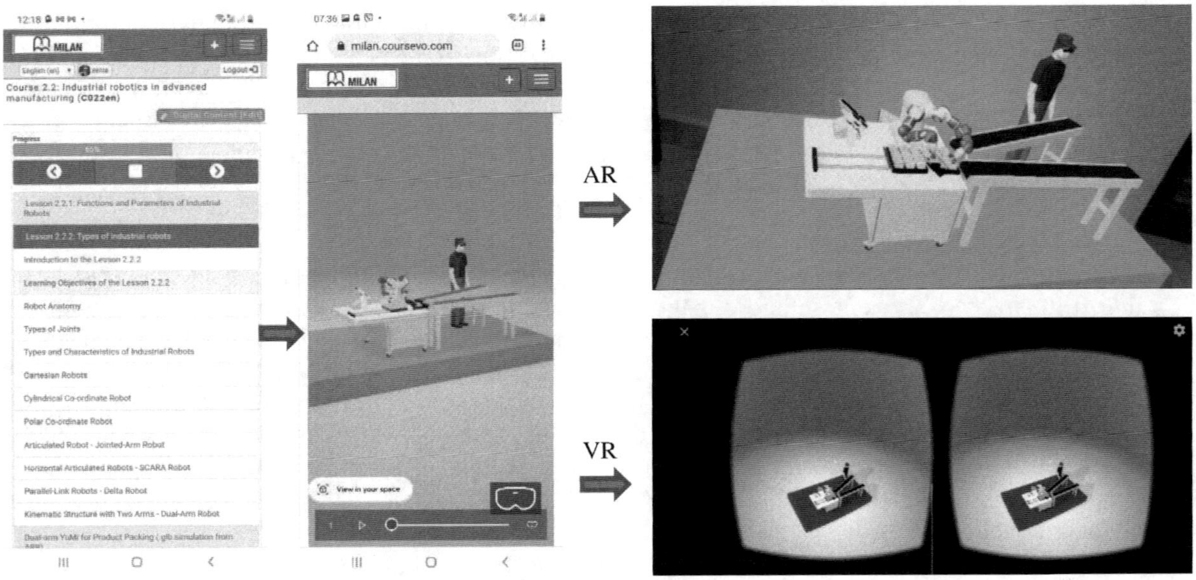

Figure 4. Integration of 3D objects and animated simulations into lessons and presentation on the Web, in AR or VR (an animated glb from ABB has been used in this example).

Figure 5. Applications developed in Unity for MILAN project (left – AR, right –VR supporting WebXR).

However, a 3D object in glb format, animated or not, cannot support alone complex AR/VR scenarios and user interactions, although it can be part of them. To build such scenarios, like interactive simulations, an application should be developed. There are several options to develop such an application. One of them is using a game engine, such as Unity, and the appropriate packages / plugins for developing and building the AR or VR applications for different devices.

In Figure 5, two applications are presented, developed in Unity for the needs of MILAN project. The first one (left part of Figure 5) is an AR application for smartphones. The student can place a robot in a real environment, control its arms by changing different parameters, and see the results. In this case, the CAD model of the IRB 1600 robot (ABB) has been used as an input, downloaded from the company site and properly edited and prepared for the needs of the application. Similarly, other robots can be used as an input, when their CAD models are available. These could also be new custom models, e.g. models of kinematic pairs, kinematic chains or even entire robot manipulators. The second one (right part of Figure 5) is a VR application, showing work of two robots taking parts from two transportation lines. Using appropriate extensions in Unity, the application implements the WebXR Device API [4], providing access to I/O capabilities commonly associated with AR and VR devices, to support immersive AR and VR experiences on the web across a wide variety of hardware. Such WebXR applications can be imported to Coursevo, integrated into learning activities, and supporting AR/VR experiences to any XR platform with a WebXR enabled web browser.

6. Summary and conclusions

In this paper we presented the MILAN project approach and training infrastructure integrating state-of-the art ICT technologies and solutions that can be effective to implement attractive forms of vocational education and training, and at the same time contribute to reducing the exclusion of people with difficult access to traditional forms of education / training - distance, transport problems, financial restrictions. MILAN supports a more flexible and accessible form of practical training, which is crucial in Robotics and Automation field, traditionally organized by robot manufacturers in physical equipment with the use of specially prepared educational stations. Towards this end, MILAN provides learning experiences supported by multimedia and AR/VR technologies, as well as advanced communication and collaboration tools allowing connections with remote laboratories.

Targeting on accessibility, MILAN exploits web-based technologies and formats, including recent standards for supporting AR/VR experiences on the Web, making its training platform and content easily accessible across a wide range of devices, not only computer stations with expensive hardware and installations but also popular mobile devices, e.g. smartphones.

Those characteristics make the MILAN approach and training platform promising in supporting training in the field of Robotics and Automation, as well as other related fields where practical training is important, in a wide audience.

The final results of the project will be available in Fall 2021. The MILAN training platform and content will be accessible in 5 languages: English, Polish, Greek, Estonian and Slovakian.

Acknowledgements

The paper presents results of research supported by EU within the project MILAN „Multifunctional Innovative Learning Assisting Network for VET in Advanced Manufacturing", 2018-1-PL01-KA202-050812, under the ERASMUS+ Programme. This publication represents only authors' opinion and neither the European Commission nor the National Agency is responsible for any of the information contained in it.

References

[1] Falkowski P, Pilat Z, Arapi P, Tamre M, Dulencin P, Homza J and Hajduk M 2021 The Concept of Using AR and VR Technologies in the Vocational Training System in Robotics and Automation *Robotics in Education. RiE 2020* ed W Lepuschitz et al. *Advances in Intelligent Systems and Computing* **vol 1316** (Cham: Springer) pp 318-325.
[2] MILAN project Homepage, http://milan-project.eu/, last accessed 2021/04/21
[3] Pappas N, Arapi P, Moumoutzis N and Christodoulakis S 2017 Supporting Learning Communities and Communities of Practice with Coursevo *Proc. IEEE Global Engineering Education Conference (EDUCON 2017)* April 2017 Athens Greece pp 297-306.
[4] WebXR Device API, https://immersive-web.github.io/webxr/, accessed 2021/05/11
[5] Khronos Group. GL Transmission Format (glTF) Specification v2.0, https://github.com/KhronosGroup/glTF/blob/master/specification/2.0/README.md, accessed 2021/03/12.
[6] Google, Using Scene Viewer to display interactive 3D models in AR from an Android app or browser, https://developers.google.com/ar/develop/java/scene-viewer, accessed 2021/04/10.
[7] Babylon.js viewer, https://doc.babylonjs.com/extensions/babylonViewer, accessed 2021/04/10.

IOP Conf. Series: Materials Science and Engineering 1239 (2022) 012002 doi:10.1088/1757-899X/1239/1/012002

Finite Element Model Updating Approach for Structural Health Monitoring of Lightweight Structures Using Surface Response Optimization

D Eidukynas[1], A Adumitroaie[2], P Griškevičius[2], V Grigaliūnas[1] and T Vaitkūnas[2]

[1] Kaunas University of Technology, Institute of Mechatronics, Kaunas, Lithuania
[2] Kaunas University of Technology, Faculty of Mechanical Engineering and Design, Kaunas, Lithuania

Abstract. Mechanical defects in the structure changes it's vibration response. There is a wide variety of methods that examine changes in measured vibration response to detect, locate, and characterize damage in structural and mechanical systems. One method to evaluate the structural changes and to analyse their causes is the Finite Element Model Updating (FEMU). The objective of this research is to investigate the FEMU procedure for mechanical damage identification and to propose an experimental-computational SHM method for lightweight structures. The structural dynamic response to impact excitation of a structure with and without defects are collected from transient and modal analysis using Ansys FE software. Afterwards, FEMU algorithm using Ansys Surface Response Optimization is investigated for its applicability to damage identification. Obtained results revealed the possibility to use this algorithm with having minimum discrepancy between parameters obtained from experiments and finite element modelling.

1. Introduction

Structural health monitoring (SHM) is tracking static or dynamic characteristics of a structure to identify and localize damage, monitor its evolution, and decides inspection and repair intervals in order to avoid the structural collapse. Mechanical changes caused by defects in mechanical structure changes it's vibration response. There are wide variety of methods that examine changes in measured vibration response to detect, locate, and characterize defects in structural and mechanical systems [1-5]. The basic idea behind this technology is that modal parameters (notably frequencies, mode shapes, and modal damping) are functions of the physical properties of the structure (mass, damping, and stiffness) [6-8]. Therefore, changes in the physical properties will cause detectable changes in the modal properties. One of the method to evaluate this changes and to find what causes these changes is Finite Element Model Updating (FEMU), which could be used in wide variety of applications: Computational FEMU algorithms for damage identification based on experimental measurements of the structural dynamic response [9], as applied to multi-response structural parameters estimation for bridges; Convolutional Neural Network based FEMU approach for the prediction of various types of damage in composite laminates based on low frequency structural vibration outputs [10]; applied to damage detection in helicopter blade based on experimental data from laser scanning vibrometery [11]; applied to structural damage detection based on measured time domain vibration response [5]; FEMU feasibility applied for the suspension bridges assessment [12]. In most of the research FEMU process is carried out using specific Matlab programming codes [13], classical mathematical calculations using FEM matrixes or Ansys APDL codes [12]. In this research a novel procedure for FEMU using Ansys FE software combined with Ansys Surface Response Optimization is proposed and investigated for the defect identification in aluminium plate. Proposed procedure could be used in significantly complex structures for SHM purposed in lightweight composites structures such as wind turbine blades, aeroplane wing, automobile frames, etc.

Content from this work may be used under the terms of the Creative Commons Attribution 3.0 licence. Any further distribution of this work must maintain attribution to the author(s) and the title of the work, journal citation and DOI.
Published under licence by IOP Publishing Ltd

2. FEMU Algorithm using Ansys Response Surface optimization software

Principal scheme of proposed FEMU algorithm is presented in Fig. 1.

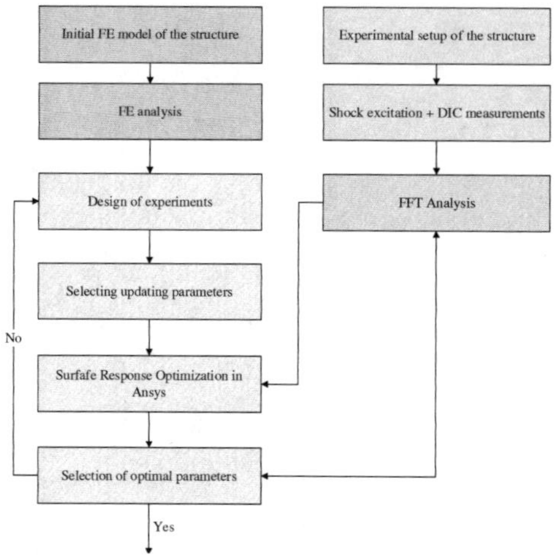

Figure 1. Principal scheme of FEMU algorithm

As it is shown in Fig. 1, at the beginning initial finite element model of the structure is created and finite element analysis in order to obtain natural frequencies, mode shapes or other required parameters is carried out. In parallel experimental setup of the same structure and contact, i.e. accelerometers, or non-contact measuring technology, i.e. digital image correlation (DIC), 3D laser vibrometry, etc. with impact type excitation or FFT analysis on demand of obtained results should be carried out. These results are compared with results obtained from FE analysis and if there are some shifts, i.e. caused by defects, between frequencies or other parameters, surface response optimization by selecting updating parameters using design of experiments tool is carried out. This procedure is repeated by iterations until difference between results obtained from experiments and FE analysis are minimized. Afterwards parameters, which were used in optimization and caused minimum difference between experimental results and FE modeling are used to identify and describe defect in the structure. For validation of proposed method for damage identification, two different cases using virtual experiments with aluminium plate with and without defects were carried out and presented in next section.

3. Damage identification using proposed FEMU algorithm

Finite element model updating procedure were carried out using Ansys R19.2 software in combination with response surface optimization tool. FE analysis was carried out with Transient and Modal analysis.

3.1. Finite element modelling of the investigated structure

FEMU in this research was conducted with two different types (stages) of virtual experiments and data sets. In the first stage Modal analysis of the plate were carried out with the health and damaged aluminum plate (two different locations of defects, i.e., 2 cases) having same geometric conditions. Defect was modelled as 20x20 mm through hole on the structure and its location varied depending on simulation case. Geometry of the investigated plate are presented in Fig. 2 and geometric dimensions and properties of the material together with FEM numerical simulation data of both stages – modal and transient analysis are collected and presented in table 1.

During both modal and transient analysis aluminium plate was fully fixed on the one surface area of 150 x 150 mm size (white zone in Fig. 2). During modal analysis 20 resonant frequencies with damaged and undamaged plate were calculated. During transient analysis excitation was applied by adding 10 N force for 0.022 s on the excitation surface *A* thus generating excitation impulse, presented as green zone in Fig. 2. Noticed, that there could be 15 different excitation surfaces *A-M* in this structure for further

IOP Conf. Series: Materials Science and Engineering 1239 (2022) 012002 doi:10.1088/1757-899X/1239/1/012002

investigation. Transient analysis results were vertical displacement of measurement points 1-24, measured for 3 s period from the excitation. Noticed, that these measurements points can be the same used for real experimental investigation, e.g. DIC or laser vibrometry measurements.

Figure 2. Geometry of the investigated aluminium plate: left side – undamaged structure, middle – with defect Case 1, right side – with defect Case 2.

Table 1. Properties, geometric dimensions of the aluminium plate and FEM numerical simulation data

Parameter	Unit	Value
length x width x thickness	mm	500 x 150 x 3
Young's modulus of aluminium alloy	GPa	71
Poisson's ratio of aluminium alloy	-	0.33
Density of aluminium alloy	kg/m^3	2770
number of finite elements without defect	-	3000
number of nodal points without defect	-	6262
number of finite elements with defect	-	2984
number of nodal points with defect	-	6244
excitation impulse time for transient analysis	s	0.022
excitation impulse for transient analysis	N	10
fixture surface area length x width	mm	150x150

Modal analysis results – natural frequencies of modes 1-8 with and without defect in the aluminium plate is presented in table 2.

Table 2. Modal analysis results

Mode number	Natural frequency, undamaged, Hz	Natural frequency, defect case 1, Hz	Difference between undamaged and case 1, Hz	Natural frequency, defect case 2, Hz	Difference between undamaged and case 2, Hz
1	20.5	20.64	0.14	20.50	0.00
2	99.54	99.80	0.27	99.04	-0.49
3	127.77	127.23	-0.54	126.69	-1.08
4	319.76	317.40	-2.36	318.43	-1.33
5	359.33	355.84	-3.49	358.53	-0.80
6	600.11	597.23	-2.88	598.34	-1.77
7	697.29	693.04	-4.25	691.59	-5.70

Mode number	Natural frequency, undamaged, Hz	Natural frequency, defect case 1, Hz	Difference between undamaged and case 1, Hz	Natural frequency, defect case 2, Hz	Difference between undamaged and case 2, Hz
8	766.16	753.61	-12.55	757.09	-9.07

As it is seen from presented results, damage decreases the natural frequency of the structure and significant differences between undamaged and damaged structures begins from mode no. 4, what means that in FEMU process is better to use higher modes of the investigates structure.
Transient analysis results, as for example measurement point no. 1 displacement versus time in 3s period is presented in Fig. 3, and measurement points 1-12 minimum vertical displacement, which as numerical value will be used for FEMU process, in the same time period is presented in table 3.

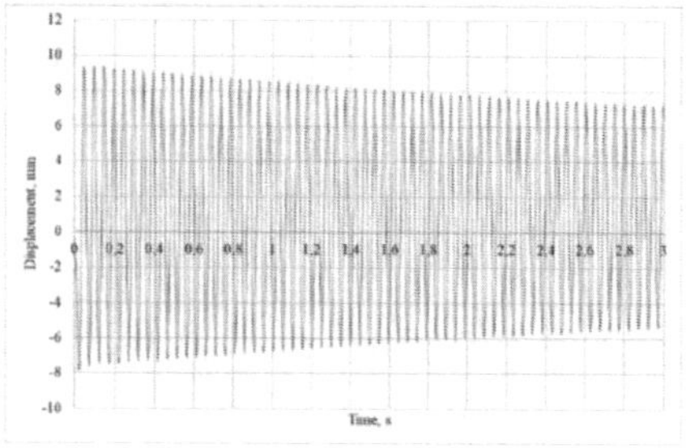

Figure 3. Measurement point no. 1 displacement in 3s period of undamaged structure

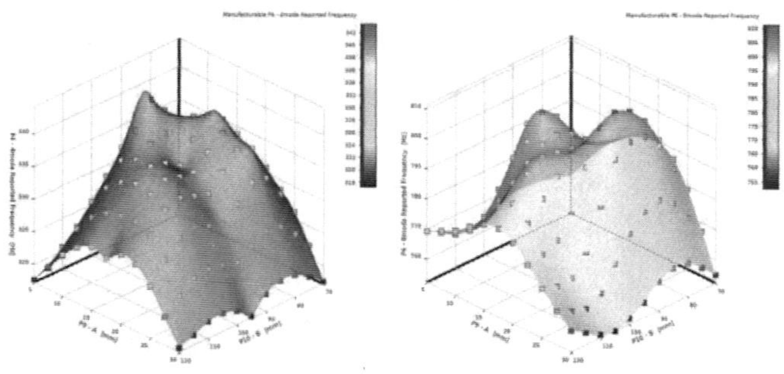

Figure 4. Response surfaces of the 4th and 8th modes for case 1

Table 3. Transient analysis results

Measurement point number	Minimum vertical displacement without defect, mm	Minimum vertical displacement defect case 1, mm	Difference between undamaged and case 1, mm	Error, %
1	7.8638	7.9228	0.0590	0.75
2	7.8027	7.8585	0.0558	0.72
3	7.7252	7.7778	0.0526	0.68
4	7.6410	7.6900	0.0490	0.64
5	5.4503	5.4700	0.0197	0.36
6	5.4998	5.5281	0.0283	0.51
7	5.5568	5.5917	0.0349	0.63
8	5.5922	5.6258	0.0336	0.60
9	3.4757	3.4901	0.0144	0.41
10	3.4761	3.4937	0.0176	0.51
11	3.4593	3.4680	0.0087	0.25
12	3.4353	3.4280	0.0073	0.21

As it is seen from presented results, damage change values of displacement, thus changing structure's response to the same excitation conditions. Noticed, that the most relevant results are up to 20th point, because points 21-24 are almost fixed.

3.2. Finite element modelling updating

In this section two different FEMU process is demonstrated to validate proposed algorithm, presented in Fig. 2: using modal and transient analysis results for FE modelling and virtual experiments results as parameters required for updating process.

FEMU process using Modal analysis data: The central composite design approach was used to perform the design of experiments. Horizontal distance of damage from the right corner and vertical distance of damage from the bottom were used as updating parameters. The interval of horizontal distance varied

from 5 to 30 mm and vertical – from 70 to 120 mm for case 1. For case 2 horizontal distance varied from 50 to 80 mm and vertical – from 140 to 190 mm. Updating values were divided into 13 and 11 parts (by incremental size equal to 2,5 and 5 mm) of horizontal and vertical distance respectively for both cases. A total of 17 design points as centre point were created and calculated. The natural frequencies of the 1st, 4th, 7th and 8th modes were selected as output parameters.
The genetic aggregation approach is used to generate response surface. Fig. 4 presents response surface as example for the 4th and 8th modes of the case 1.
After creating response surface, optimization using surface response optimization tool and Moga optimization method was carried out. For this purpose, natural frequencies obtained from virtual experiments, presented in the 3rd and 5th column in table 2 were used as objective values for algorithm to seek this target. Table 4 presents the optimization of design variables for horizontal and vertical distance and natural frequencies of 2 candidate points for case 1 and case 2.

Table 4. Optimization results for FEMU using modal analysis

Parameter		Virtual experiment		Candidate 1		Error, %		Candidate 2		Error, %	
		Case 1	Case 2	Case 1	Case 2	Case 1	Case 2	Case 1	Case 2	Case 1	Case 2
Defect location, mm	Horizontal distance	30	55	30	50	0.00	3.45	5	140	0.00	3.45
	Vertical distance	70	145	70	140	0.00	9.09	70	80	83.33	45.45
Natural frequencies, Hz	Mode 1	20.64	20.5	20.61	20.51	0.15	0.04	20.64	20.41	0.00	0.44
	Mode 4	317.40	318.43	317.62	318.27	0.07	0.05	317.42	318.30	0.01	0.04
	Mode 7	693.04	691.59	693.64	692.14	0.09	0.08	696.60	690.60	0.51	0.14
	Mode 8	753.61	757.09	753.31	757.21	0.04	0.02	764.13	756.29	1.40	0.11
Average						0.06	2.12			14.21	8.27

As it is seen for the obtained results, candidate no 1 fits case 1 with average error 0.06 % and has error factor 2.12 % for case 2. However, to measure higher modes of the investigated structure might be a challenge, thus similar procedure was tested using transient analysis results and modelling thus imitating excitation of the first mode only and measuring displacement of the surface when excitation condition is known and remains the same for all tests.

FEMU process using Transient analysis data: The initial updating parameters and intervals were the same as described in 3.2.1 chapter for case 1. The minimum vertical displacement in 3s time of measurement points 1, 6, 11 and 12 together with the 1st natural frequency of the structure were selected as output parameters. For optimization minimum vertical displacement of measurement points 1, 6, 11 and 12 obtained from virtual experiments, presented in the 3rd column in table 3 were used as objective values for algorithm to seek this target. Table 5 presents the optimization using transient analysis results of 3 candidate points for case 1.

Table 5. Optimization results for FEMU using transient analysis

Parameter		Virtual experiment	Candidate 1	Error, %	Candidate 2	Error, %	Candidate 3	Error, %
Defect location, mm	Horizontal distance	30	29.5	1.67	30	0.00	5	83.33
	Vertical distance	70	69.8	0.29	82	17.14	78	11.43
Natural frequency, Hz	Mode 1	20.64	20.63	0.05	20.623	0.08	20.621	0.09
Minimum vertical displacement, mm	Point 1	7.9228	7.9221	0.01	7.924	0.02	7.936	0.17
	Point 6	5.5281	5.5275	0.01	5.5284	0.01	5.5287	0.01
	Point 11	3.468	3.46	0.23	3.4674	0.02	3.4696	0.05
	Point 12	3.428	3.423	0.15	3.4251	0.08	3.4275	0.01

Parameter	Virtual experiment	Candidate 1	Error, %	Candidate 2	Error, %	Candidate 3	Error, %
Average			0.34		2.48		13.58

Obtained results showed that the candidate 1 fits optimization scope and thus identifies defect with average error up to 0.34 %. Candidate no 2 can be considered as well due to error up to 2.48% while candidate no 3 is not relevant in this case.

4. Conclusions

Finite element modelling updating technique using surface response optimization were proposed and investigated. Obtained results showed that using this method it is possible to identify and locate damage by using two different measurement and data acquisition techniques. It is necessary to measure at least 4 natural frequencies of modes 1, 4, 7 and 8 of investigated structure to identify defect with accuracy up to 99.94%. Since measurements of higher modes natural frequencies are complicated in real application, the methodology was tested with measuring surface displacement and the first natural frequency only. Obtained results revealed the possibility to identify defect in the structure with accuracy up to 99.66 %. Noticed, that this algorithm can be applied with various updating and objective parameters, such as elastic modulus, stiffness, damping ratio, etc.

5. References

[1] Seguel F.and Meruane V 2018 Damage assessment in a sandwich panel based on full-field vibration measurements. *Journal of Sound and Vibration*, 417, 1-18.

[2] Zhang Z, Pan J, Luo W, Ramakrishnan K R and Singh H K 2019 Vibration-based delamination detection in curved composite plates. *Composites Part A: Applied Science and Manufacturing*, 119, 261-274.

[3] Martinez-Luengo M, Kolios A and Wang L 2016 Structural health monitoring of offshore wind turbines: A review through the Statistical Pattern Recognition Paradigm. *Renewable and Sustainable Energy Reviews*, 64, 91-105.

[4] Carden E P and Fanning P 2004 Vibration based condition monitoring: a review. *Structural health monitoring*, 3(4), 355-377.

[5] Garcia D, Palazzetti R, Trendafilova I, Fiorini C and Zucchelli A 2015 Vibration-based delamination diagnosis and modelling for composite laminate plates. *Composite Structures*, 130, 155-162.

[6] Gomes G F, de Almeida F A, Junqueira D M, da Cunha Jr S S and Ancelotti Jr A C 2019 Optimized damage identification in CFRP plates by reduced mode shapes and GA-ANN methods. *Engineering Structures*, 181, 111-123.

[7] Amadori S and Catania G 2017 Robust identification of the mechanical properties of viscoelastic non standard materials by means of frequency domain experimental measurements. *Composite Structures*, 169, 79-89.

[8] Umar S, Bakhary N and Abidin A R Z 2018 Response surface methodology for damage detection using frequency and mode shape. *Measurement*, 115, 258-268.

[9] Sanayei M, Khaloo A, Gul M and Catbas F N 2015 Automated finite element model updating of a scale bridge model using measured static and modal test data. *Engineering Structures*, 102, 66-79.

[10] Khan A, Ko D K, Lim S C and Kim H S 2019 Structural vibration-based classification and prediction of delamination in smart composite laminates using deep learning neural network. *Composites Part B: Engineering*, 161, 586-594.

[11] Link M, Weiland M and Seckert T 2008 Computational model updating for damage identification in the time domain. *Proceedings of ISMA 2008.*

[12] Merce R N, Doz G N, de Brito J L V, Macdonald J H and Friswell M I 2007 Finite element model updating of a suspension bridge using ansys software. *In Design and Optimization Symposium.*

[13] Maljaars P J, Kaminski M L and Den Besten J H 2017 Finite element modelling and model updating of small scale composite propellers. *Composite Structures*, 176, 154-163.

Acknowledgments
This research is supported by Research Council of Lithuania (Project CompExSHM No.: P-MIP-19-523).

IOP Conf. Series: Materials Science and Engineering 1239 (2022) 012003 doi:10.1088/1757-899X/1239/1/012003

Investigation of the resistance, durability and abrasion of printed elements applied to sports equipment

A Sidorov, I Iljin

Department of Mechatronics, Robotics and Digital Manufacturing, Faculty of Mechanics, Vilnius Gediminas Technical University [Saulėtekio al. 11, LT-10223] Vilnius, Lithuania

Abstract. The work investigated the durability of the labelling of sports equipment made of composite materials. Professional hockey sticks were taken for research. Research has been done on hockey stick wear for different players: centre and winger and defender. In addition, a study was made of hockey sticks for left-handers and right-handers. In the study of the wear of printed elements, sports equipment of various lengths was used. Contact research was carried out during training and competition. The study of the number of contacts was carried out on the basis of video recordings of trainings and games of the teams "Energija" - Elektrėnai (Lithuania) and the national team of the Belarusian Republic. The study carried out showed that the hockey stick wear of the centre striker and the winger during the competition and training is similar. During the competition, the hockey sticks of the centre strikers are most often in contact. It was found that the difference between contacts during training and playing depends on the intensity of the game, penalty time and strategy. When conducting a study of the wear of the printed elements applied to the hockey stick, depending on the position of the player, it was found that the wear of the sticks of the winger and the defender is the same, but that of the centre striker is different. When conducting a study of hockey stick wear depending on its length, the following conclusion was made: the taller the player, the longer his hockey stick and the maximum contact area. Research has shown that hockey stick wear is not affected by how the player holds the hockey stick, i.e. whether he is left-handed or right-handed. To study the abrasion of elements using the UGRA method, an original stand was created. With the help of this stand, studies of the abrasion of printed elements were carried out. Based on the study, it was concluded that the abrasion of printed elements most often occurs at the moment of contact of the objects of research, both with each other and with objects on the playing field. As a result, the printed elements split, peel off and become less noticeable, in addition, the name of the manufacturer's company is worn out and the hockey stick loses the player's personal identification data printed on the hockey stick. Recommendations are given on the peculiarities of printing and the location of printed elements on sports equipment.

1. Introduction

The production of sports equipment is constantly being improved, manufacturers use more and more new materials, introduce new technologies, and the choice of hockey sticks is huge. Speaking of composite hockey sticks [1], we must understand that we are not talking about a cheap product, the purchase of a hockey sticks today is becoming more and more individual. Modern hockey sticks made from composite materials are well balanced in weight.

There are thickening zones, rubberized zones and so on the stick. If earlier the handle of the hockey sticks was bent equally along the entire length, today the bend is asymmetric. The hockey sticks has recently changed its geometry. If earlier the handle was uniform along the entire length, today we have

Content from this work may be used under the terms of the Creative Commons Attribution 3.0 licence. Any further distribution of this work must maintain attribution to the author(s) and the title of the work, journal citation and DOI.
Published under licence by IOP Publishing Ltd

a new model of hockey sticks, which are pointed towards the bottom. The closer the handle is to the hook, the thinner it becomes and thus stiffer. This design allows for better puck control and stronger, more accurate shots.
To identify the hockey stick by manufacturer, team and player, various inscriptions are applied to it.
For the manufacture of a hollow stick handle, in order to achieve uniform rigidity of the product, 15 ... 20 sheets of carbon fiber material are superimposed on each other in different directions, the layers are pressed together with a thermoactive resin. The base for the hook is a plastic core. In the center there is an insert made of foam rubber, pasted over with carbon fiber, which absorbs the impact of the puck.
The aim of the work is to study the durability of the durability of markings (inscriptions, individual data of the player), hockey sticks. For this, professional sports equipment of the firms "Bauer", "True", "CCM" was selected, which were made with the use of various design elements, the study of abrasion of which was carried out using a stand developed according to the UGRA method. Based on the recorded games and training sessions, the frequency of hockey stick contact with various objects was calculated.

2. Materials and methods

The object of the test is the inscriptions on the handle of the stick, over which a protective silicone coating is applied. The identification of the presented samples was made on the basis of manufacturers and the location of the players on the field. The names of the tested samples, the data on the manufacturer and the player's position are presented in table 1:

Table 1. Hockey positions on the court and the manufacturer of the stick

Player position	Test sample of a hockey stick	Number of ice trips for one period	Number of stick contacts with objects per game	The number of contacts with objects in one exit	Time of going out on the ice, for the period, s	Number of contacts during training (1 hour)
Defender	Bauer Vapor	4…9	20…30	3…5	30…50	15…20
Winger	Bauer Vapor	4…7	30…45	3…6	30…50	10…25
Center forward	Bauer Vapor	4…7	30…45	5…9	30…50	10…25
Defender	TRUE	3…8	25…40	3…7	30…60	15…25
Winger	TRUE	4…6	30…50	4…8	30…50	10…20
Center forward	TRUE	4…7	30…60	5…12	30…55	10…35
Defender	CCM	3…9	20…45	3…5	30…45	10…25
Winger	CCM	4…8	30…50	3…6	30…60	10…30
Center forward	CCM	4…7	30…40	4…9	30…50	10…25

When identifying the preparation of the samples submitted for testing, an actual comparison of the main characteristics of the surface of the samples under study with new products was carried out. The products presented for testing have different markings depending on the manufacturer, however, the outer silicone coating is even, uniform, matte and transparent, without smudges and pores. The assessment of the condition of the coating was carried out in comparison with new control hockey sticks. The material of the products is Kevlar K49, a para-aramid fiber manufactured by DuPont [2]
The tests were carried out according to the following scheme: a) the number of contacts during training was investigated, b) the number of contacts during the game was investigated, c) accelerated tests were carried out using the stand and according to UGRA [3]. Results of field tests, which were carried out on the basis of video recordings of the games of the Eletrenai - Energija team of the 2019-2020 season.

The study involved a sample of 30 hockey sticks from three major manufacturers: Bauer Vapor, TRUE, CCM, 10 from each. No specific hockey stick model has been emphasized. The number of defects after the games was calculated, taking as the main condition that before the game all the hockey sticks were new, without defects. Visual assessment of the condition of the coating during the test was carried out according to ISO 4628: 2016 [4, 5]

3. Results and discussion

During visual inspection of the condition, the general appearance of the new product was assessed (Figure. 1a) and re-examination for the presence of defects after play, characterizing the deterioration of the decorative properties of the product. Surface defects are cracking, flaking, discoloration, gloss, abrasion of print elements (Figure. 1b). This is necessary to further predict the life of the product.

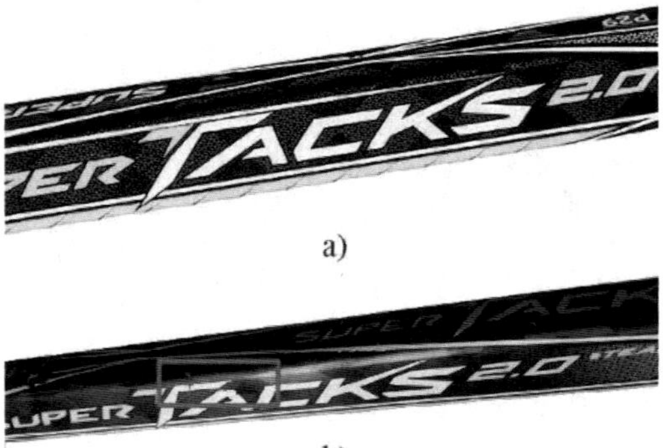

a)

b)

Figure 1. Condition of the CCM Super Tacks 2.0 Stick Review hockey stick before the game (a) and after the game (b).

Adhesion determines the quality of adhesion of the printing coating to the surface of the composite material from which the stick is made. This is a necessary property, in the absence of which the product must be rejected and replaced with a new one, since it is the undamaged image that is the main indicator of the surface quality. Poor adhesion can be caused by poor degreasing of the Kevlar surface, improper coating by the manufacturer, or poor quality paints incompatible with composite materials [6].
Poor quality of paint and varnish coating on the composite material of the product after playing indicates poor adhesion of paints and a protective layer with the composite material of the product body [7].
To detect contact damage (Figure 2 a), a measurement scheme was proposed, shown in Figure 2b. The stick is tilted at an angle of 60° relative to the horizontal plane, that is, as is usually the case in the game.

IOP Conf. Series: Materials Science and Engineering 1239 (2022) 012003 doi:10.1088/1757-899X/1239/1/012003

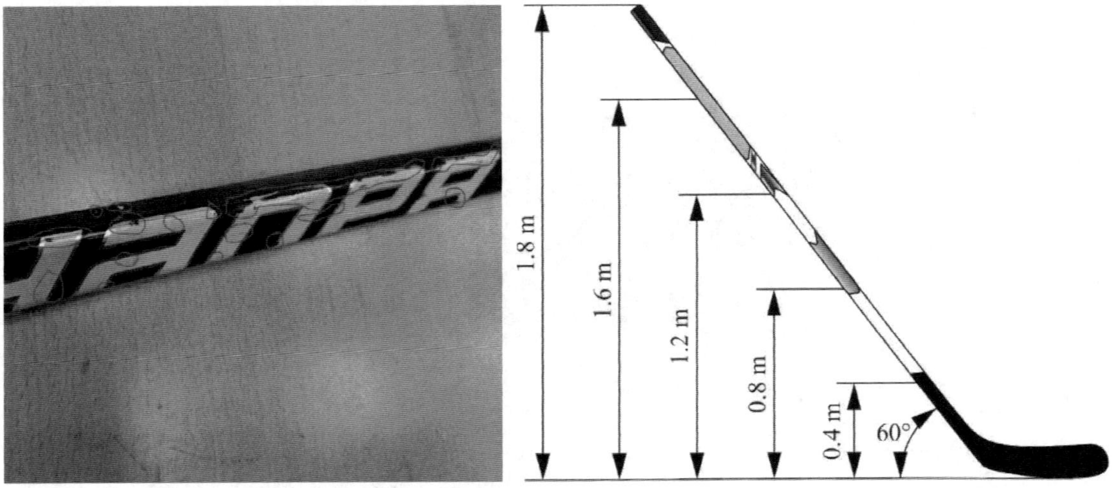

a) b)

Figure 2. Places for measuring contact damage.

During the experiment during the competition and training, the hockey stick was worn out. During one exit on the ice, the contact of the winger's stick with other objects occurs from 5 to 15 times in one period (Figure. 3-1). During the game, sports equipment contacts other objects (sticks, goal structures, puck, protective glazing structures, etc.) up to 35 ... 45 times, which is clearly visible on the video.
It should be noted that the defender's play is less intense than the striker's play, however, there is also an intense wear of the sports equipment (Figure. 3-2). As a result, in one exit on the ice, the hockey stick contacts other objects from 3 ... 5 to 15 times in one period. Contact with other objects occurs up to 20 ... 30 times.
It was revealed that the play of the central forward is the most intense (Figure. 3-3), and one way out on the ice, the stick contacts other objects from 5 to 15 times per period and up to 10 times during contact serve, which is 30..50 times per the game.

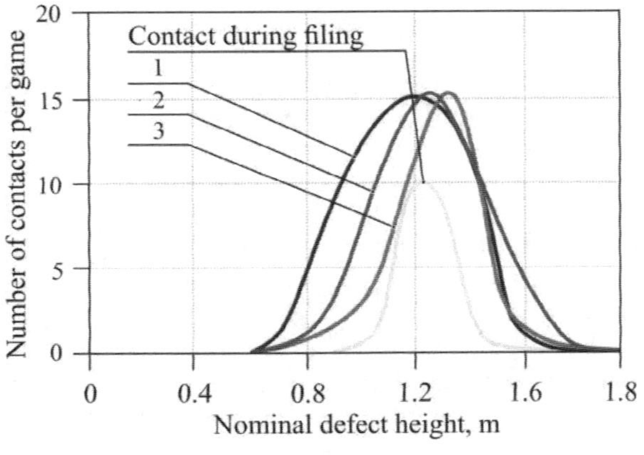

Figure 3. Distribution of stick contact injuries: 1- winger, 2- defender, 3 - center forward.

To accelerate the study of the abrasion of printed elements, a study was carried out on the abrasion stand of the UGRA tester [UGRA, 2017]. The device consists of three elements: a guide carriage with a stand, a sliding sled for abrasion, a loose metal block in a sled (weight 1639 g), on which is applied artificial leather similar to that used on hockey gloves (figure 4). The dimensions of the metal block are 107x64 mm. The block causes a pressure of 0.5 N / cm2. The friction path is 2x10 cm per double movement.

IOP Conf. Series: Materials Science and Engineering 1239 (2022) 012003 doi:10.1088/1757-899X/1239/1/012003

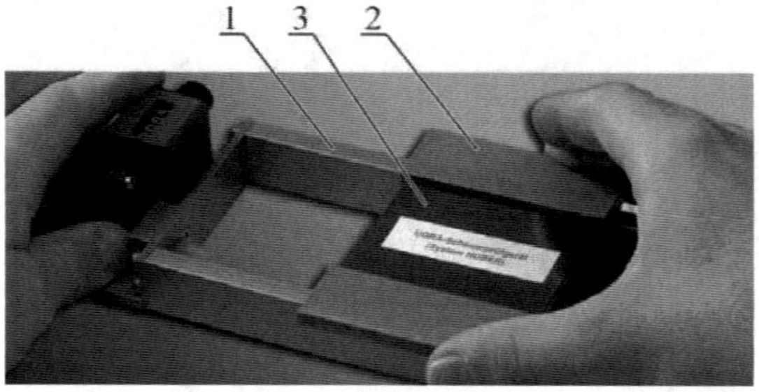

Figure 4. UGRA PCB Abrasion Tester

In the course of the experiment, it was found that the complete abrasion of the protective layer of silicone and printed elements occurs after 700 cycles. The degree of surface abrasion was estimated as a percentage from 0 to 100%, so that the complete abrasion of the image and the protective coating corresponded to 100% (Figure 5).

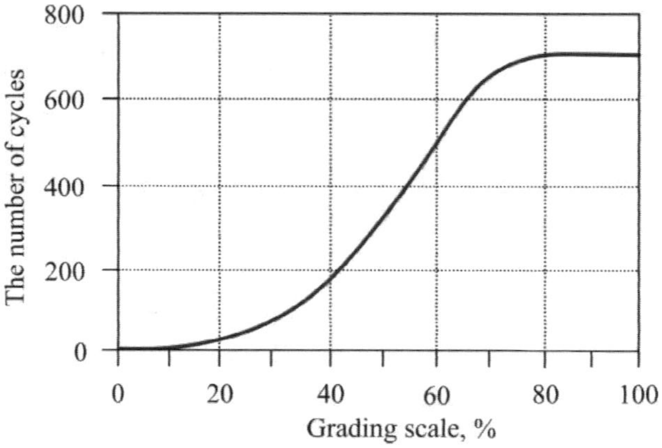

Figure 5. Printed element abrasion graph

4. Conclusions:

1. After analyzing the play of the defender, winger and center striker during the game and training, it was found that the wear of the hockey sticks at the moment of contact is similar, but during the most often the hockey stick of the center striker, as the main player in the team.

2. After completing the study of the video, it was found that the center forward during the throw-in of the puck by the referee has frequent contact with the stick of the opposing team's player. The study of video footage of the games showed that the center striker has 10 ... 20 more shots than the rest of the players.

3. Having completed the research, it was revealed that the hockey stick wear is uneven and the main wear occurs at a height of 1.2 ... 1.3 m. In addition, wear does not depend on whether the player is right-handed or left-handed.

4. Having completed the research, it was revealed that the main wear of the hockey stick occurs during contact with the sticks of the opposing team's player, the puck and other objects on the field, as well as the fencing and glazing of the court. In this case, the printed elements break off, peel off and, due to

IOP Conf. Series: Materials Science and Engineering 1239 (2022) 012003 doi:10.1088/1757-899X/1239/1/012003

abrasion of the outer layer, become less noticeable, which in turn spoils the outer one. In addition, such important elements as the name, surname and number of the player are damaged.

5. Wear of a hockey stick from a player's gloves is less relevant than contact damage during the game and occurs over a greater number of cycles.

6. Based on the above, it is recommended to replace the hockey stick before each, in addition, due to the high cost of the product, it is recommended to restore the outer layer by grinding it to a depth of 0.01 ... 0.05 mm, followed by gluing printed applications ordered from the manufacturer, or by applying inscriptions on site by screen printing with the final coating of the product with a protective silicone layer.

5. References:

[1] Bauer. 2021 Team unity (2021).
[2] DuPont, Technical Guide for Kevlar Aramid Fiber: (2017).
[3] UGRA. Ugra Abrasion Tester (2017).
[4] ISO 4628:2016 Paints and varnishes — Evaluation of degradation of coatings — Designation of quantity and size of defects, and of intensity of uniform changes in appearance. Part 4: Assessment of degree of cracking (2016).
[5] ISO 4628:2016 Paints and varnishes — Evaluation of degradation of coatings — Designation of quantity and size of defects, and of intensity of uniform changes in appearance. Part 5: Assessment of degree of flaking (2016).
[6] H. P. Le, Progress and Trends in Ink-jet Printing Technology, Journal of Imaging Science and Technology 1(42) (1998)
[7] H. Kipphan, Handbook of Print Media. Springer – Verlag: Berlin-Heidelberg-New York. (2001).

Experimental research of the damping characteristics of the fatigue stand caused by changes in material properties

R Pawliczek[1,2], **T Lagoda**[1] **M Kurek**[1] **and G Robak**[1]

[1] Faculty of Mechanical Engineering, Opole University of Technology, Mikolajczyka st. 5, 45-271, Opole, Poland
[2] corresponding author

Abstract. The paper presents the results of research on changes in the damping characteristics of the MZGS100 fatigue test stand. The work of the stand is influenced by the type of material of the tested samples, where different properties affect the dynamic characteristics of the stand. It is important to assess the impact of changes in properties during the test on the work of the test stand. Fatigue tests were carried out for three materials: S355J0 steel 2017A-T4 aluminium alloy and CuZn40Pb2 brass. The tests were carried out under fatigue bending loads for the number of failure cycles in the low-cycle range. During the investigation, the test was interrupted and the impulse response of the stand at various stages was recorded and the change in system response was assessed. It was observed that in the case of brass, the damping curve of the system changed continuously, and the damping factor of the system increased by 23%. In the case of steel, practically no changes were noticed, while the aluminium alloy showed changes of 9%. In the final stage of the test for steel and aluminium alloy, a change in the frequency of natural vibrations was observed, which indicates a decrease in the system stiffness caused by the weakening of the specimen`s cross-section. This effect was not noticed for brass samples, which together with the change of the damping coefficient indicates a cyclical change of the material properties.

1. Introduction

The phenomenon of material fatigue is a well-known factor that destroys structures that operate under variable loads. Standard fatigue characteristics of materials obtained in laboratory tests are used in the analysis and design of structures in terms of material fatigue. Such tests are usually based on standards and the machines are required to carry out the tests under controlled conditions of loading or deformation of the test pieces. For this reason, modern test stands are equipped with measuring elements and actuators, which, in cooperation with controllers with implemented control programs, constitute modern machine solutions, representing a mechatronic approach to machine construction [1, 2]. Such a system makes it possible to control the response of the material during the test and to react to any changes in the material caused by the phenomena associated with material fatigue.

Construction materials are characterized by different properties, which means that the work of the test stand depends on the material being tested. The rigidity of the machine-sample system causes the conditions for the operation of the control system to change. Basic relations of load and deformation in the material due to the different resistance of the material to fatigue loads have an influence on the dynamics of the device [3]. Differences in material properties affect the system response frequency and damping.

In fatigue tests of materials, the effect of changing the properties of the material subjected to fatigue loads for the so-called cyclically unstable materials, is well known [4, 5]. It is important to assess the impact of changes in properties during the test on the work of the test stand. For such behavior of

Content from this work may be used under the terms of the Creative Commons Attribution 3.0 licence. Any further distribution of this work must maintain attribution to the author(s) and the title of the work, journal citation and DOI.
Published under licence by IOP Publishing Ltd

IOP Conf. Series: Materials Science and Engineering 1239 (2022) 012004 doi:10.1088/1757-899X/1239/1/012004

materials, it is necessary to recognize machine operation from the point of view of automatic control systems. The type of the system and its parameters must be recognized and it must be decided whether this will have an influence on the operation of the machine.

The aim of the paper is to investigate the system response of the fatigue stand MZGS100 due to changes in material properties during the test.

2. Description of the MZGS100 fatigue test stand

This paper presents operation of the MZGS100 fatigue test stand (Figure 1), which was developed for fatigue tests of specimens made from constructional materials subjected to combined bending and torsional loading [6].

Figure 1. The view and the scheme of the MZGS100 stand [6].

Operation of the stand is based on inertial vibro-mechanism. Vibrations of the rotating disk are generated by the inertial force with amplitude equal to centrifugal force *B* of the unbalanced weights (Figure 2). Then, the vertical displacement is transmitted to the load lever by link. A material specimen is loaded with the cyclic main moment M(t)

$$M(t) = M_a \cdot sin(\omega_w t) = B \cdot l \cdot sin(\omega_w t), \tag{1}$$

where: M_m – mean value of the load, M_a – amplitude of the load, ω_w – periodicity of the rotating disk. The structure of the stand represents typical form of second order oscillatory system. Assuming the displacement y(t) as the output of the system its operation can be described by equation:

$$\ddot{y} + 2\varsigma\omega\dot{y} + \omega^2 y = B \cdot sin(\omega_w t), \tag{2}$$

where: ς – damping ratio, ω – natural periodicity of the system.
Parameters of the system can be defined experimentally base on impulse response. A mathematical form of the equation representing impulse response for second order oscillatory system is

$$y(t) = k\left[\frac{\omega}{\sqrt{1-\varsigma^2}} e^{-\varsigma\omega t} \cdot sin(\omega\sqrt{1-\varsigma^2}\, t)\right]. \tag{3}$$

By detecting the peaks of the impulse response (Figure 2), one can find the damping curve and determine its parameters by approximation using the function:

$$y(t) = ae^{-bt} + c, \tag{4}$$

where: *a* - amplitude, *b* – decay rate, *c*- offset. After normalizing and removing the trend in the signal, a = 1 and c = 0 are obtained. By comparing equations (3) and (4) it can be written:

$$b = \varsigma\omega \qquad 2\pi f = \omega\sqrt{1-\varsigma^2}, \tag{5}$$

where *f* – frequency of the impulse response signal. Parameters *b* and *f* can be obtained experimentally and in equation (5) we have two unknowns: ς and ω.

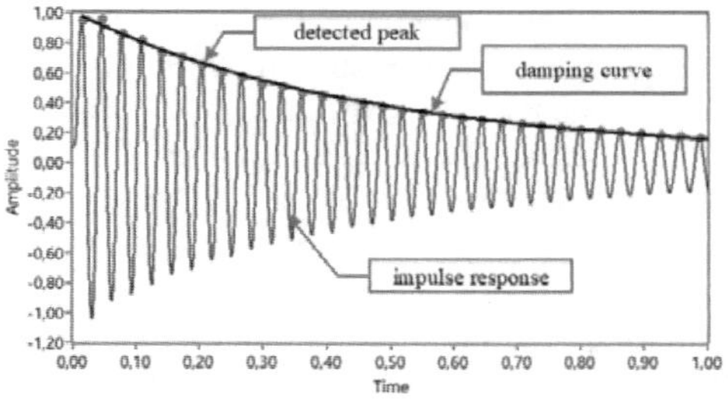

Figure 2. Impulse response.

The only replaceable element in the system is the sample, as well as its shape and dimensions as well as the type of material. Therefore, the natural frequency of the stand depends on the sample. Assuming the sample as articulated beam (does not transmit moment), the natural frequency of the stand is f_n = 27 Hz.

3. The test procedure and materials

The aim of the research is to determine changes in the parameters of the impulse response along with the number of load change cycles. If the material properties change during the test, it will be possible to observe their influence on the parameters of the impulse response. The impulse response of the system was recorded for each material sample before starting the tests, and then the fatigue test was launched. After the specified number of cycles was completed, the test was stopped and the impulse response was re-examined. In this way, system responses were obtained for several stages of the specimen life.

For each recorded impulse response, the trend was removed and then the signal was normalized with respect to the maximum value of the displacement. For the recorded time waveforms, the value of the damping coefficient b and the frequency of the impulse response f in Hz were determined. Then, on the basis of Eq. (5), the damping factor ς and the natural periodicity ω were determined.

The shape and dimensions of the specimens are presented in Figure 4 and it was the same for all of the tested materials. The materials used for the tests are S355J0 steel, 2017A-T4 aluminium alloy and CuZn40Pb2 brass. The basic strength properties of the materials are presented in Table 1.

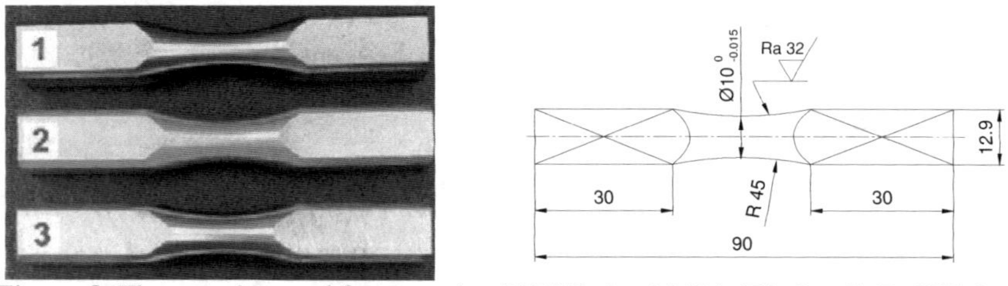

Figure 3. The samples used for tests: 1 – S355J0, 2 – 2017A-T4, 3 – CuZn40Pb2

Table 1. Material strength properties.

	Young`s modulus E *GPa*	Tensile ultimate strength σ_U *MPa*	Yield strength limit σ_y *MPa*	Relative elongation A5 *%*
S355J0	213	611	394	21
2017A-T4	72	545	395	21
CuZn40Pb2	96	484	400	10

4. The test results and analysis

Sample results of impulse response measurements for specimens before running the test are shown in Figure 4a. The damping curves for this cases are also presented in the Figure 4b. The strongest damping in the system occurs for brass CuZN40Pb2, while the 2017A-T4 aluminium alloy shows the weakest damping from the analysed cases.

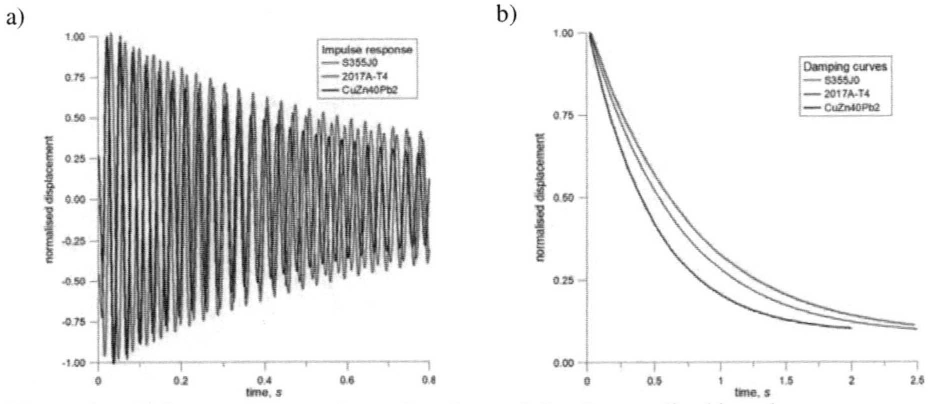

Figure 4. Initial state of system for analysed materials: a) normalised impulse response, b) damping curves

In order to compare the results of the analysis during the fatigue tests, the relative test stage coefficient was used in the form of n_i/N_f, where n_i is the current number of cycles and N_f is the number of cycles to failure of the sample, e.g. the value of the coefficient $n_i/N_f = 0.5$ means that the impulse response test was performed halfway through the test. The values of coefficient b in Eq.(4) for normalised impulse response investigations during the tests are presented in Figure 5.

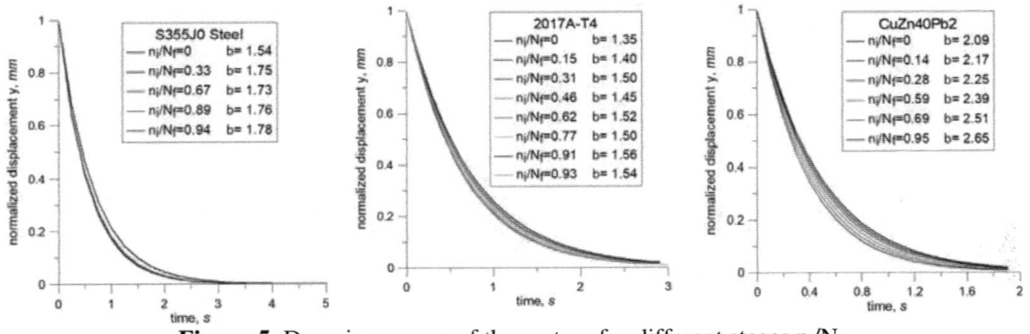

Figure 5. Damping curves of the system for different stages n_i/N_f.

It was observed that after the initial weakening for all materials, S355J0 steel shows a stabilization of the coefficient b at the level of about 1.77. For the aluminium alloy, a change in the value of b between 1.40 and 1.56 was observed without an unequivocal trend. On the other hand, for brass, a systematic increase in the coefficient b was observed with the number of cycles n and from 2.17 to 2.65 - a change of 22%. In the case of brass, we can talk about a continuous change of material properties, which results in such behaviour of the system.

Table 2 presents the results of calculations of the damping factor ς and the natural periodicity ω for the tested cases, and Figure 6 shows a graphical presentation of the results.

It was noticed that in the case of S355J0 steel and 2017A-T4 aluminium alloy in the final stage of the test (results marked with a rectangle in Figure 6), there was a large decrease in the system's natural frequency f_n. This is due to the initiation and propagation of fatigue cracks. During this time, the active cross-section of the sample is significantly reduced, causing a very large drop in the stiffness of the system, which is manifested mainly in the reduction of the natural frequency. At the same time, a slight change in the damping factor ς is observed. In the case of CuZn40Pb2 brass, the natural frequency does not change practically until the end of the test, where the material rapidly plasticizes and the displacement y(t) increases. At the same time, a continuous increase in the value of the damping factor ς is observed. These changes are greater than for steel and aluminium alloy.

Table 2. Parameters of the mathematical model of the system.

Material	test stage n_i/N_f	damping ratio ς	natural periodicity ω, *1/s*	natural frequency $f_n=2\pi/\omega$, *Hz*
S355J0	0	0.0078	198.6	31.6
	0.33	0.0088	198.6	31.6
	0.67	0.0087	198.6	31.6
	0.89	0.0093	188.5	30.0
	0.94	0.0106	168.4	26.8
2017A-T4	0	0.0073	186.0	29.6
	0.15	0.0075	186.0	29.6
	0.31	0.0081	186.0	29.6
	0.46	0.0078	186.0	29.6
	0.62	0.0082	186.0	29.6
	0.77	0.0081	186.0	29.6
	0.91	0.0087	178.4	28.4
	0.93	0.0090	170.9	27.2
CuZn40Pb2	0	0.0104	201.1	32.0
	0.14	0.0108	201.1	32.0
	0.28	0.0112	201.1	32.0
	0.59	0.0119	201.1	32.0
	0.69	0.0125	201.1	32.0
	0.95	0.0133	198.6	31.6

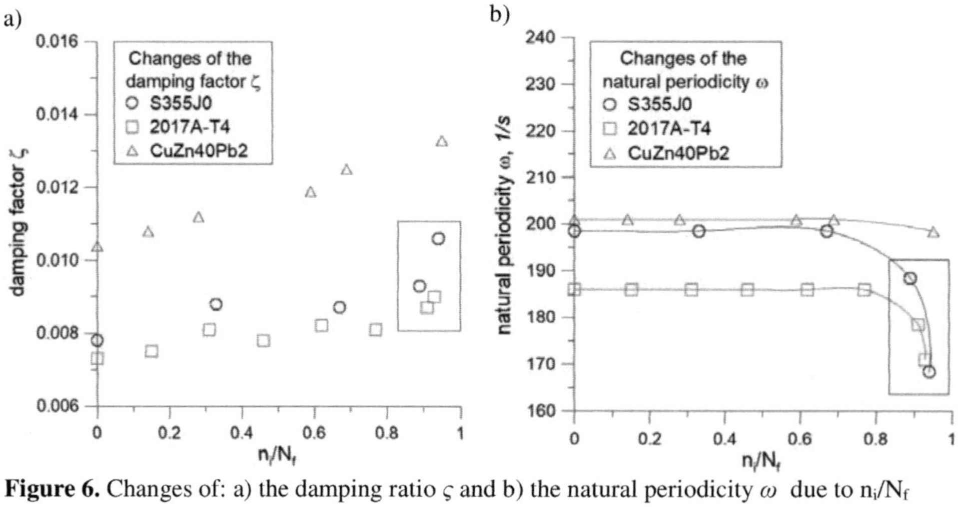

Figure 6. Changes of: a) the damping ratio ς and b) the natural periodicity ω due to n_i/N_f stage.

4. Summary and conclusions

The stable operation of the MZGS100 test stand depends primarily on the properties of the tested material. The sample is an integral part of the system and affects its characteristics. The work presents fatigue tests for steel, aluminium and brass alloys, where the impulse response of the system at various stages of the tests was recorded and analysed during the test. Based on the test results and data analysis, the following conclusions can be made:

- for the initial state of the material the strongest damping in the system was recorded for a sample made of CuZn40Pb2 brass. In the case of S355J0 steel 20117A-T4 of the aluminium alloy, the damping in the system differed by about 14%, while brass showed about 36% higher damping effect of the system in relation to steel.
- in the final stage of the test (n_i/N_f> 0.9) for steel and aluminium alloy, a sudden decrease in the value of the natural frequency f_n was observed, with the frequency value tending to f = 27 Hz, which is the natural frequency of the stand with the articulated beam sample. This means that at this stage there is already a failure in the form of a crack and the stiffness decreases rapidly.
- a constant increase in the damping ratio ς was observed (by 23%) with the increasing number of load cycles for brass, for aluminium alloy these changes were 9% and 2% respectively (excluding the final stage of the test).
- in the case of bronze CuZn40Pb2, the natural frequency of the system remains practically until the end of the test, while the damping ratio changes continuously. Such behaviour proves a systematic change of material properties at every stage of the research.

References

[1] Kabaciński M and Pawliczek R 2015 Mechatronic concept for airflow test laboratory equipment, *Solid State Phenomena* Vol. 220-221 pp. 445-50

[2] Pawliczek R 2020 Modernization of the fatigue test stand control system using the idea of a virtual instrument, 15th Int. Conf. Mechatronic Systems and Materials (Bialystok, Poland), *IEEE Xplore*, 2020, pp. 1-6

[3] Lagoda T, Kurek M, Robak G, and Pawliczek R 2017 Influence of the specimen's material and shape on behavior of the fatigue test stand MZGS100 *The 13th Int. Conf. Mechatronic Systems and Materials (MSM – 2017) – Abstract*, pp. 69-70

[4] Mrozinski S and Piotrowski M 2016 Effect of Strain Level on Cyclic Properties of S355 Steel, *Fatigue Failure And Fracture Mechanics XXVI, AIP Conf. Proc.*, 020005 1780

[5] Kurek M, Kurek A and Lagoda T 2019 Stress-life curve for high and low cycle fatigue, *Journal of Theoretical and Applied Mechanics* 57(3) pp. 677-84

[6] Achtelik H, Kurek M, Kurek A, Kluger K, Pawliczek R and Lagoda T 2018 Non-standard fatigue stands for material testing under bending and torsion loadings, *AIP Conf. Proc.* 2029 020001

IOP Conf. Series: Materials Science and Engineering 1239 (2022) 012005 doi:10.1088/1757-899X/1239/1/012005

CFD simulation of heat transfer and airflows in an open refrigerated display cabinet

Tadas Vengalis* **and Vadim Mokšin**

Department of Mechanical and Materials Engineering, Vilnius Gediminas Technical University, J. Basanavičiaus str. 28, LT-03224 Vilnius, Lithuania

*Corresponding author e-mail: tadas.vengalis@vilniustech.lt

Abstract. This paper uses CFD (Computational Fluid Dynamics) method to study the efficiency of the airflow curtain in an empty and full open refrigerated display cabinet. Numerical simulation results show that changing of the angle of the honeycomb can reduce the infiltration of warm ambient air into the refrigerator. It was established that the average intake air temperature in modified design decreased by 0.6 °C compared to the original design. The CFD model presented in the paper can be used to study airflows inside and to design and develop open-type refrigerated cabinets.

Keywords: Open refrigerated display cabinet; CFD; Heat transfer; Airflow; Refrigeration; Temperature; Air velocity

1. Introduction

Most foods sold in supermarkets require cooling. Energy-efficient refrigerated cabinets are needed to keep food fresh. In order to increase the sales, food sellers usually use open-type refrigerators. These refrigerators are less energy efficient compared to closed-type refrigerators, but the market dictates its terms. To improve the performance of open-type refrigerators, single or double airflow curtains are used, which act as the door. The air inside the refrigerator is separated as much as possible from the warm ambient air and human actions. Refrigerated display cabinets must maintain the set temperature in accordance with [1]. For most foods like milk, dairy products, fried meat products and others, 3M2 class refrigerators are required that maintain the temperature of the products in the range from –1 to +7 °C. This temperature range is easily achievable in the most open-type refrigerated cabinets. To keep fresh meat in display cases, 3M0 class refrigerators are required, which maintain the temperature of foodstuffs in the interval from –1 to +4 °C. This temperature interval is rarely achieved in open-type refrigerators due to high ambient air infiltration rates inside the refrigerator. To improve the performance of the refrigerator, it is necessary to study the work of the airflow curtain and improve it. Minor geometric changes can sufficiently improve the efficiency of open-type refrigerated cabinets. However, designing of an energy-efficient open-type refrigerator requires a lot of laboratory testing to measure the infiltration of warm and humid ambient air into the interior of the refrigerator, that can make this task economically difficult for many refrigerator manufacturers.

Computational fluid dynamics (CFD) simulation software can be applied to reduce the cost and amount of laboratory testing of refrigerators. Many studies have shown that CFD simulation is a promising tool to study refrigeration efficiency and temperature mode in the refrigeration zone. Wang

Content from this work may be used under the terms of the Creative Commons Attribution 3.0 licence. Any further distribution of this work must maintain attribution to the author(s) and the title of the work, journal citation and DOI.
Published under licence by IOP Publishing Ltd

IOP Conf. Series: Materials Science and Engineering 1239 (2022) 012005 doi:10.1088/1757-899X/1239/1/012005

et al. [2] reported the progress in 3D simulation of a full refrigeration cabinet system under dynamic operation of automated switching on and off. Sun *et al.* [3] investigated the impact of flow guiding strips on the flow of the air curtain, the temperature of the products and (most importantly) on the energy consumption of the cabinet. Chaomuang *et al.* [4] developed a simplified heat transfer model of a closed display cabinet based on the zonal approach. This model enables the identification of the critical zones, i. e., too warm and too cold, which can lead to food quality deterioration and chilling injury, respectively. Rai *et al.* [5] conducted multiple CFD simulations to study the changes in thermal flow pattern inside a refrigerated truck body with and without an air curtain.

This paper presents a steady-state 2D numerical analysis of airflow curtains and temperature changes in an open refrigerated cabinet at different airflow honeycomb inclination angles. The heat transfer and airflows were simulated with commercial CFD software COMSOL Multiphysics 5.5. The aim of this article is to study the influence of the honeycomb angle on the formed air curtain and temperature in refrigerated cabinet. Minor changes in cabinet geometry can reduce the refrigeration temperature, thus improving the refrigeration class.

2. CFD model of refrigeration cabinet

2.1. Research object

The open-type vertical refrigerated display cabinet is schematically shown in figure 1. This cabinet is used for food products that should be stored at temperatures between –1 and +7 °C. It has 5 shelves for food products, their dimensions are shown in figure 1. The air enters into the cabinet through the air grill at the front of the bottom panel. Fans blow air through the evaporator. Then cooled air travels through a tunnel to the top of the refrigerator. The perforated distributor at the top distributes the cooled air. Part of the air is blown through the perforated back panel and the other part passes through the air-off honeycomb at the front top of the cabinet, thus forming air curtain between the inside of the refrigerator and the ambient warm air to protect the chilled food products. The cooled air entering through the back perforated panel into the display area helps to maintain the required food temperature.

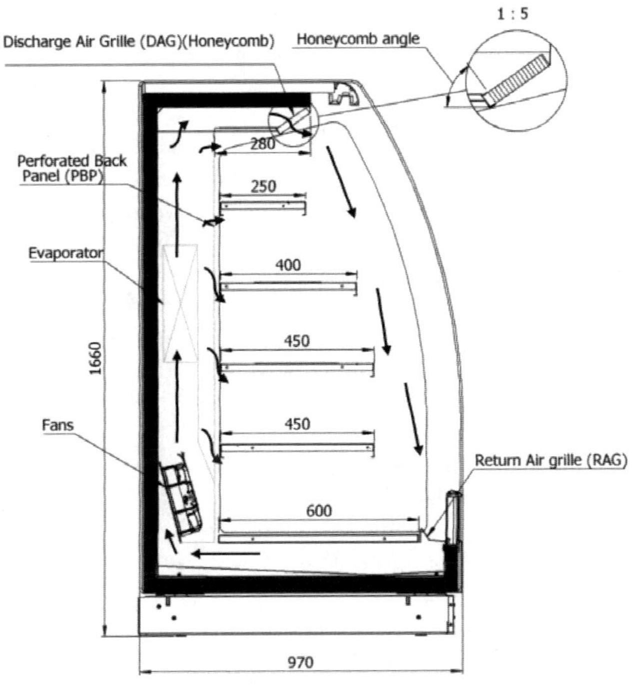

Figure 1. Side view of the refrigerated cabinet.

The overall dimensions of the cabinet are as follows: 1660 mm in height, 970 mm in depth and 2500 mm in width. Spacing between base and first shelf is 280 mm, between other shelves – 250 mm. The air-off honeycomb dimensions are as follows: 20 mm (height) × 120 mm (width). Its inclination angle (figure 1) with respect to the horizontal plane was varied from 40 to 65°. All model dimensions including back panel, shelves, air ducts, etc. were the same as dimensions of real cabinet.

2.2. Mesh and simulation conditions

"COMSOL Multiphysics" software was used to simulate airflows and heat transfer in the refrigerated cabinet. Figure 2 shows the mesh of 2D model of the cabinet. The total number of 2D finite elements is about 12 000. A higher number of finite elements does not influence the results significantly.

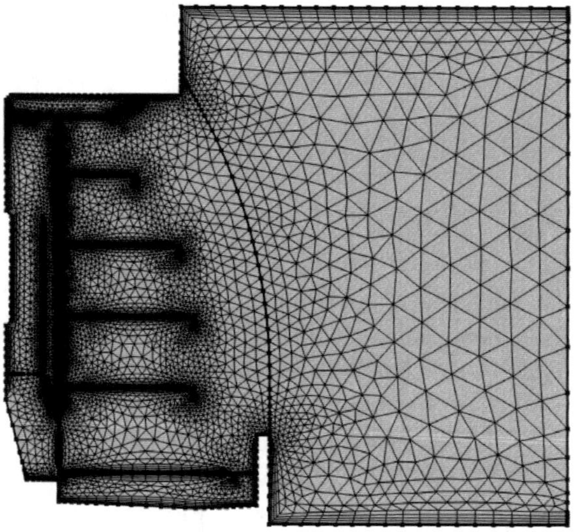

Figure 2. 2D mesh of the air space inside and outside of the refrigerated cabinet.

In the COMSOL Multiphysics model, the fan was simulated with normal outflow velocity, which value was set as 1.2 m/s. The outflow air temperature was considered as constant –1 °C value. This value was measured above the evaporator in the experimental study.

The open-type refrigerated cabinet exhibits strong natural convection between chilled air curtain and warm ambient air. To simulate this natural convection, the gravity was activated in the model and the density of the air was calculated based on the ideal gas formula.

In this study, the thin layer structure was used for honeycomb modelling to provide more accurate flow direction. Hanged shelves, bottom shelf, back perforated panel and other parts of the cabinet are made from a sheet of steel through which heat transfer occurs. Hanged shelves and bottom shelf were modelled as steel walls with a thickness of 1.4 mm. Other inner parts were modelled as steel walls with a thickness of 0.7 mm.

According to the standard [1], a refrigerated display cabinet should be tested in a controlled environment at 25 °C, 60 % RH and 0.1–0.2 m/s airflow velocity conditions for climate class 3. For simulation open boundary condition was applied to the model. Open boundary condition means that air can freely move in and out of the testing environment.

3. Simulation results and discussion

Figure 3 presents profiles of a fully formed airflow curtain at the middle plane of the empty refrigerated cabinet obtained for the different honeycomb angles. In all cases, the airflow curtain fully formed after 60 seconds. It can be seen from figure 3 that as the honeycomb angle increases, the curtain straightens, airflow radius increases. For 55–65° angle interval, the airflow lines almost do not cross the refrigerated cabinet outline presented by black curve in figure 3. The velocity of the airflow out of the honeycomb is 0.6 m/s and it decreases to 0.4 m/s near the intake grill at the bottom of the cabinet. The reason for reduced airflow curtain velocity is cold and warm air mixing in all curtain length.

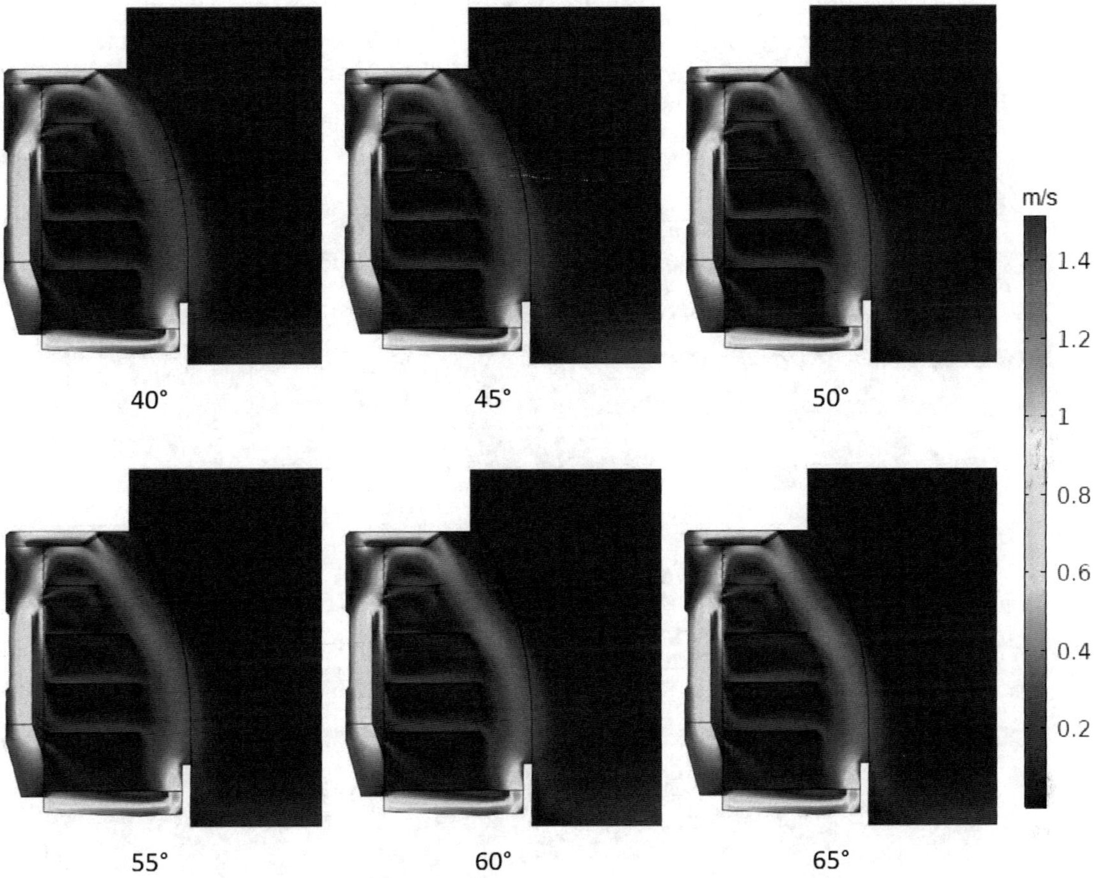

Figure 3. Airflow velocity distribution inside the empty refrigerated cabinet obtained for different honeycomb angles.

Figure 4 presents an intake air temperature versus honeycomb angle graph obtained for case of empty refrigerated cabinet. It can be seen from figure 4 that the honeycomb angle influences the temperature of the intake air. In case of 55° angle the temperature is lowest that means that this airflow angle is optimal for case of empty cabinet.

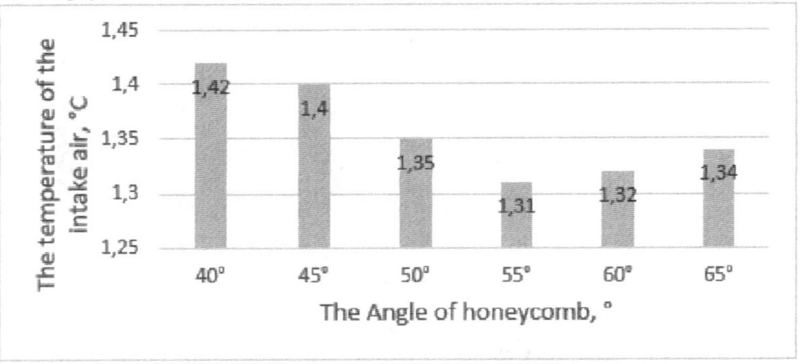

Figure 4. Air temperature at the intake grill versus honeycomb angle (empty cabinet).

Figure 5 presents profiles of a fully formed airflow curtain at the middle plane of the full refrigerated cabinet obtained for the different honeycomb angles. In the case of full (with food products) cabinet, the airflow curtain fully formed after 120 s. However, the airflow curtain changes its shape compared with empty cabinet case (figure 3) and its lines cross the refrigerated cabinet outline over the entire range of honeycomb angles. For 40–45° angle interval, the airflow is formed outside the cabinet outline. In case of 55–65° angle interval, the airflow curtain is formed at the outline and environmental factors do not influence its shape sufficiently. Velocities of the airflow out of the honeycomb and near the grill are the same as obtained for case of empty cabinet (figure 3).

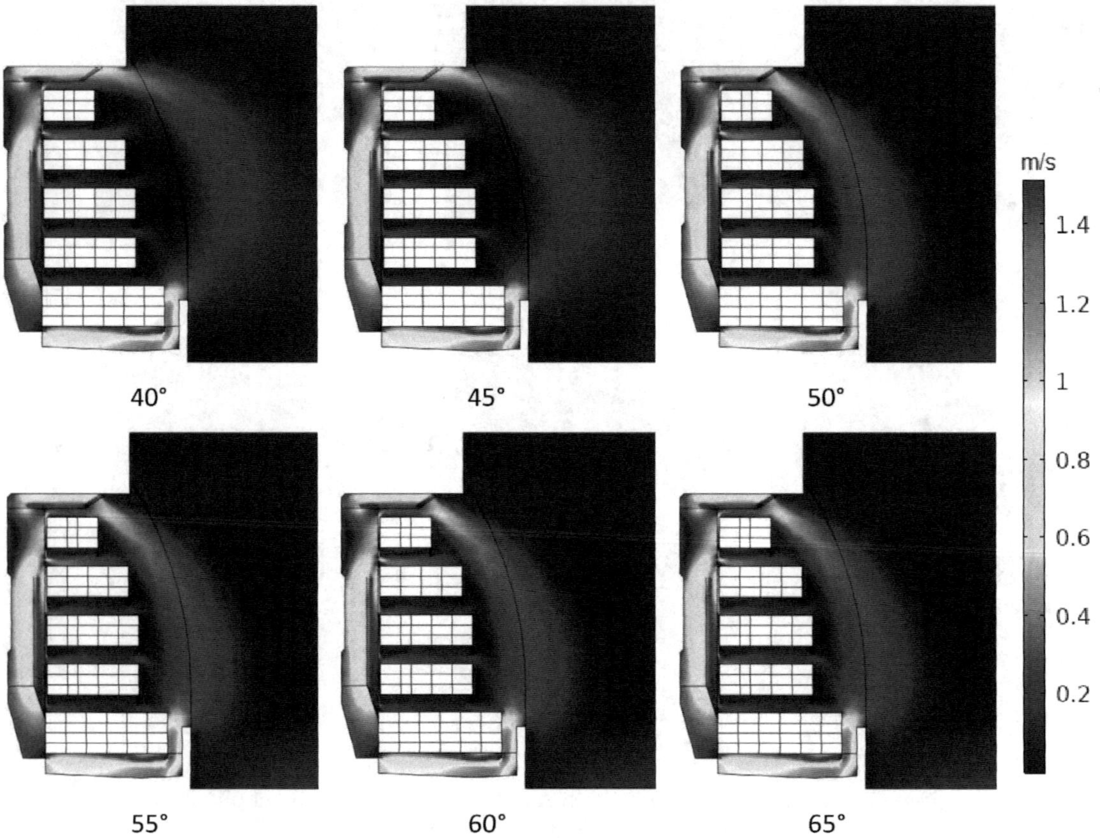

Figure 5. Airflow velocity distribution inside the full refrigerated cabinet obtained for different honeycomb angles.

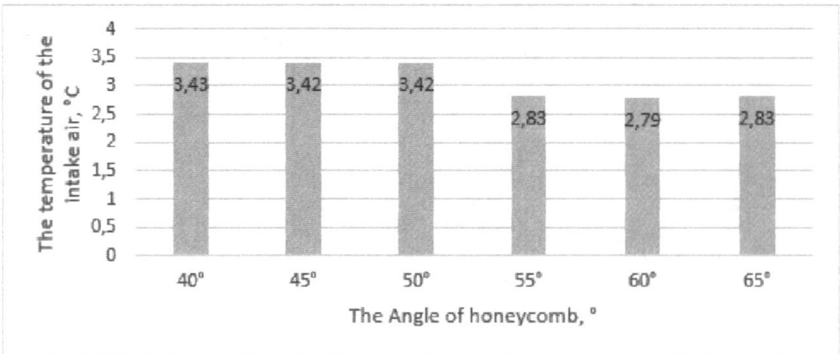

Figure 6. Air temperature at the intake grill versus honeycomb angle (full cabinet).

Figure 6 presents an intake air temperature versus honeycomb angle graph obtained for case of full refrigerated cabinet. It can be seen from figure 6 that temperature of the intake air decreases with increase of honeycomb angle. The temperature simulation results are different from results obtained for empty refrigerated cabinet (figure 4). The heat exchange between the airflow curtain and the environment takes longer because the airflow rate remains the same, but the path it travels is increased, so the elongated airflow curtain does not completely separate the interior of the refrigerator from the warm ambient air.

4. Conclusions

- A simplified steady-state heat transfer model was developed to analyze airflows and heat transfer in open-type refrigerated cabinet according with ISO 23953-2 standard.
- Numerical simulation results show that intake air temperature decreases with an increase of honeycomb angle in both cases of empty and full refrigerated cabinet.
- Numerical simulation results show that the highest efficiency of the airflow curtain is achieved by increasing of the honeycomb angle to 55° only. Further increase of the angle doesn't lead to significant changes in efficiency.
- The simple heat transfer model developed for this study will be used in further studies investigating the effect of hanging shelf lengths on food temperature. The 2D research methodology will be used to study 3D models.

References

[1] 2015 *European Standard ISO 23953-2:2015 Refrigerated Display Cabinets – Part 2: Classification, Requirements and Test Conditions* (Geneva: The International Organization for Standardization)

[2] Wang L, Zhang L and Lian G 2015 A CFD simulation of 3D air flow and temperature variation in refrigeration cabinet *Procedia Eng.* **102** 1599–1611

[3] Sun J, Tsamos K M and Tassou S A 2017 CFD comparisons of open-type refrigerated display cabinets with/without air guiding strips *Energy Procedia* **123** 54–61

[4] Chaomuang N, Laguerre O and Flick D 2020 A simplified heat transfer model of a closed refrigerated display cabinet *Therm. Sci. Eng. Prog.* **17** 100494

[5] Rai A, Sun J and Tassou S A 2019 Numerical investigation of the protective mechanisms of air curtain in a refrigerated truck during door openings *Energy Procedia* **161** 216–223

IOP Conf. Series: Materials Science and Engineering 1239 (2022) 012006 doi:10.1088/1757-899X/1239/1/012006

Gold nanoparticle-based biofuel cell catalytic efficiency reliance on medium pH

Antanas Zinovicius[1], Juste Rozene[1], Vakare Guoba Sataite[1], Timas Merkelis[2], Joris Juska [2], Inga Morkvenaite-Vilkonciene[1]

[1] Department of Mechatronics, Robotics, and Digital Manufacturing, Faculty of Mechanics, Vilnius Gediminas Technical University, 03224 Vilnius, Lithuania;
[2] Department of Physical Chemistry, Faculty of Chemistry and Geosciences, Vilnius University, 03225 Vilnius, Lithuania;

Abstract. Over the past several decades, biology, chemistry, and medicine took advantage of gold nanoparticles (AuNPs) for their fascinating characteristics. Today these nanomaterials can be synthesized in various methods from biosynthesis, when particles are made by microbial culture, to physical. Using the fast scan cyclic voltammetry, AuNPs can be electrodeposited from chloroauric acid ($HAuCl_4$) directly onto the surface of a graphite electrode. This method provides high reproducibility and allows to control the size of the nanoparticles (from 4 to 75 nm) and electrode coverage by changing $HAuCl_4$ concentration, scan speed, and a number of applied cycles. Gold nanoparticles are known to express comparable catalytic activity to natural enzymes such as glucose oxidase or horseradish peroxidase. Despite the advantages over natural enzymes, inorganic enzyme mimicking catalysts exhibit limited catalytical efficiency and selectivity. To increase efficiency, we used low size AuNPs which increase reactivity; the lack of selectivity in a biofuel cell is advantageous since AuNPs can reduce/oxidize several substrates in the medium. Biofuel cell based on AuNP's has a superior life span, as the main active part of the system is inorganic, and thus it is not affected by oxygen stress, where typical microbial biofuel cells are struggling. The observed catalytic activity had a minimal dependency on temperature change. pH was changed from 4 to 10 with the smallest synthesized AuNPs. However, pH has an impact on the reaction speed. Results showed that biofuel cell is suitable for raw wastewater as it usually deviates from pH 7, and usable current can be generated.

1. Introduction

Biofuel cells are devices that transform chemical energy into electrical energy utilizing reactions that involve biochemical pathways. Biofuel cells can be divided into two distinct groups: microbial fuel cells and enzyme-based fuel cells. Microbial fuel cells use whole living organisms to complete enzyme pathways, while enzyme-based fuel cells use isolated and purified enzymes to catalyse specific electrochemical reactions [1].

The use of biofuel cells for the generation of electrical power has several advantages. Firstly, biofuel cells are substantially less harmful to the environment than traditional batteries since they provide power from non-toxic materials that occur naturally, such as glucose. Secondly, biofuel cells

Content from this work may be used under the terms of the Creative Commons Attribution 3.0 licence. Any further distribution of this work must maintain attribution to the author(s) and the title of the work, journal citation and DOI.
Published under licence by IOP Publishing Ltd

can operate under mild conditions (20-40°C and around neutral pH), making them desirable for application where harsh conditions are unwanted. In addition, biofuel cells can consume a wide variety of fuel substances [2, 1].

Enzyme-based fuel cells are promising in being applied in implantable medical devices, including pacemakers, drug pumps, and blood glucose meters. Current implantable medical devices use lithium-ion batteries for power, which require replacement from time to time. This replacement process leads to high medical costs and causes pain to the patients. Enzyme-based biofuel cells are biocompatible and can have long lifetimes due to the fact that they use glucose in the blood for power and oxygen as an oxidating agent. They do not require recharging as the body supplies the fuel and have high efficiency. Integrating enzyme-based biofuel cells into implantable medical devices would reduce medical costs and alleviate pain for the patients [3].

However, several drawbacks prevent the practical usage of current enzyme-based biofuel cells. A considerable problem is an instability. Enzymes are commonly active in a solution at body temperature for only a few days. This can be overcome to increase enzyme stability for up to 45 days by immobilising the enzymes. Most enzyme-based biofuel cells also depend on cofactors such as NAD+ and NADP+, which poses another instability issue. Mediated biofuel cells also suffer from another layer of instability. Organometallic mediators tend to suffer from loss of their ligands even if they show reversibility over multiple voltammetric scans. Most organic mediators experience a loss of stability due to dimerisation and can lead to the deterioration of electrodes in the cell. Enzymatic biofuel cells also carry a risk of being harmful to the body as they operate close to living tissue, and bodily fluids passing by the cell may suffer from denaturation [4].

Another issue is mass transfer—the diffusion of the fuel limits the efficacy of biofuel cells to the active sites of the catalysts. When there is high resistance to the fuel transfer, a concentration difference tends to build up between the bulk phase and the active site of the catalyst, which slows down the rate of the reaction and can cause polarisation of electrodes. Lastly, the power density of biofuel cells is a big issue. Power density is usually measured by power output per surface area of the electrode. A necessary step to achieve high power density is high enzyme loading. For example, when glucose oxidase was randomly packed as a monolayer on a flat surface was only 1.7×1012 mol/cm^2, which was determined by the physical size of the enzyme. The theoretical highest current density for this layer was calculated at about 0.2 mA/cm^2. Glucose oxidase is known as one of the most efficient redox enzymes. Therefore other enzymes with lower specific activity should produce even less power [5].

Nanozymes are catalytic nanomaterials with enzyme-like characteristics. While natural enzymes are selective, highly efficient and active catalysts, due to some of their intrinsic drawbacks such as high cost, laborious preparation, degeneration, sensitivity of catalytic activity to environmental conditions and difficulty of recycling, natural enzyme practical application is limited. Nanozymes surpass natural enzymes in these terms (while also being biocompatible) as they are simply prepared, low in cost, highly stable and easily recycled [6,7]. Besides excelling in certain aspects compared to natural enzymes, nanozymes in addition exhibit unique features: their size, structure and composition can be adjusted, changing catalytic activity. Their high surface area can be easily modified and bio-conjugated, they have multiple uses besides catalysis and have a smart response to external stimuli [6]. Yet, nanozymes, unlike natural enzymes, lack in substrate selectivity and catalytic activity, leaving space for improvement [8]. Lack of substrate selectivity could be used as an advantage when constructing nanozyme-based fuel cell. It is discovered that AuNPs could also act as a glucose oxidase mimic catalytically reacting glucose with oxygen to form gluconic acid and hydrogen peroxide [9]. The reaction rate is slower at lower pH, and because one of the products of glucose oxidation is gluconic acid, therefore the reaction is self-limiting. Adding a base can compensate for the pH drop [10].

The AuNP catalysed oxidation of glucose reaction follows an Eley-Rideal mechanism. In summary, the hydrated glucose anion adsorbs to the AuNP leading to electron-rich AuNP. After that, oxygen is activated by forming a dioxogold intermediate, then two electrons are transferred from the adsorbed glucose to oxygen via the dioxogold intermediate forming H_2O_2 and gluconate anion [6].

IOP Conf. Series: Materials Science and Engineering 1239 (2022) 012006 doi:10.1088/1757-899X/1239/1/012006

The aim of this paper is to evaluate biofuel cells catalytic efficiency reliance on medium pH using gold nanoparticles as catalyst in glucose oxidation reaction.

2. Materials and Methods

2.1. Materials

The 0.05 M phosphate-acetate buffer solution (PABS) was prepared by dissolving 0.05 M CH_3COONa, 0.05 M NaH_2PO_4, 0.05 M Na_2HPO_4 in distilled water. To increase buffer solutions conductivity 0.1 M KCl was added to it. 0.1 M $HAuCl_4$ solution was prepared in 0.5 M H_2SO_4. All chemicals were purchased from Sigma-Aldrich (Steinheim, Germany). Glucose (≥98%) and potassium ferricyanide (≥99.0%) were purchased from Carl Roth (Karlsruhe, Germany). All solutions were prepared in buffer. Glucose solution was left to mutarotate overnight before use.

2.2. Graphite electrode preparation

A small piece of the graphite electrode rod (150 mm, ø3 mm, low density, 99,995% trace metal basis) was cut and sanded with 4 different grit size sandpaper, to finish polished with paper until shinny surface is achieved. Then prepared electrode is washed with 97% ethanol to remove any grease.

To deposit gold nanoparticles onto the electrode surface fast-scan cyclic voltammetry was used. Deposition was performed using three-electrode electrochemical cell. Earlier prepared graphite rod used as a working electrode, platinum wire as an auxiliary electrode and the Ag/AgCl/KCl (3M) electrode as a reference electrode. All the components (borosilicate glass titration vessel with plastic mounting ring and lid, platinum, and 12.5 cm length Ag/AgCl/KCl (3M) electrodes) were purchased from Metrohm AG (Herisau, Switzerland). The deposition of AuNPs was performed by cycling in 0.1 M $HAuCl_4$ solution at different number of cycles at 12.5 V/s scan rate, in range from 0 to +850 mV. All AuNPs were deposited at the same conditions, to form differently-sized nanoparticles number of scans changed from 10 to 500. To verify that gold was deposited onto the electrode regular cyclic voltammetry in range from -0.7 to 1.3 V, at a scan rate of 0.1 V/s was performed to observe oxidation and reduction peaks of gold and gold(III).

2.3. Electrochemical measurements

Electrochemical measurements were performed using DorpSens μ400 potenciostat (Utrecht, the Netherlands) and DropView software. All experiments were carried out in ambient temperature (at 20 ◦C) in phosphate-acetate buffer solution (pH from 4 to 10) under aerobic conditions. The cycles were measured five at a time, and the last one was plotted in the result section. If the concentration of additional materials were changed, it was done sequentially by adding it to the same measuring cell. In between, additions solution was stirred for 30 s and left to sit for 1 min. All measurements were repeated 5 times, and the average values were derived. Cyclic voltammograms were recorded at a scan rate of 0.1 V/s in the range from −0.7 V to 0.7 V.

2.4. Calculations

The Hill equation is used to determine how many ligand molecules bind to a protein macromolecule to produce a functional effect. However, the Hill coefficient gives an accurate estimation of how many ligands bind to a protein only if there is extreme cooperativity between the binding of the first and subsequent ligand molecules. Therefore, the Hill equation is sometimes used to describe cooperativity among multiple ligand binding sites, and the Hill coefficient can be thought of as an "interaction" coefficient [11].

IOP Conf. Series: Materials Science and Engineering 1239 (2022) 012006 doi:10.1088/1757-899X/1239/1/012006

Electrochemical measurements were evaluated by Hill's equation:

$$J = \frac{C^n}{k^n + C^n} \tag{1}$$

where: J is current density, C is the concentration of substrate (glucose), k is the half-maximum concentration constant, and n is the Hill coefficient.

3. Results and Discussion

Cyclic voltammograms were recorded using graphite electrode modified with 40 nm size gold nanoparticles, 750 μM of potassium ferricyanide and incrementally changing glucose concentration form 0 to 50 mM and then changing medium to fresh with different pH value and oxidation peaks were plotted as dependency on glucose concentration (Figure 1). pH value was changed from 4 to 10 by increments of one. In all samples, current density increased with the increase of glucose concentration. The most significant change was observed at 9 pH, from 90 to 125 μA/cm^2. The 9 pH could be contributing to neutralising gluconolactone and, in this way, pushing reaction towards the product side.

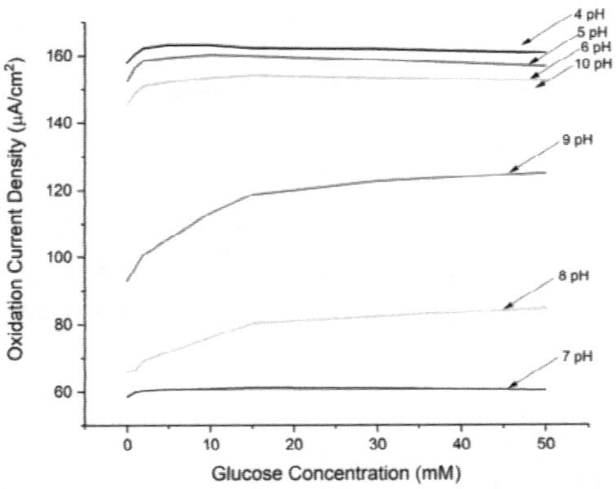

Figure 1. Current density dependency on glucose concentration when medium pH was changed from 4 to 10. The medium consists of PABS of different pH (from 4 to 10), 0.75 mM of potassium ferricyanide, and glucose.

The oxidation and reduction peaks of current density from voltammogram were plotted as glucose concentration dependence and evaluated by Hill's equation (Equation 1) (Figure 2). At the 0.15 V potential, the Hill coefficient n was higher than one (n=1.26), and at the 0.35 V potential, n was also higher than one (n=1.09), indicating positive cooperativity at both oxidation and reduction processes. If glucose concentration increases, then the efficiency of charge transfer by glucose molecules also increases. Half the maximum concentration constant k (k_{ox}=8.123) was lower at oxidation potential than at reduction potential (k_{red}=9.08), which emphasises that oxidation reaction rate was faster.

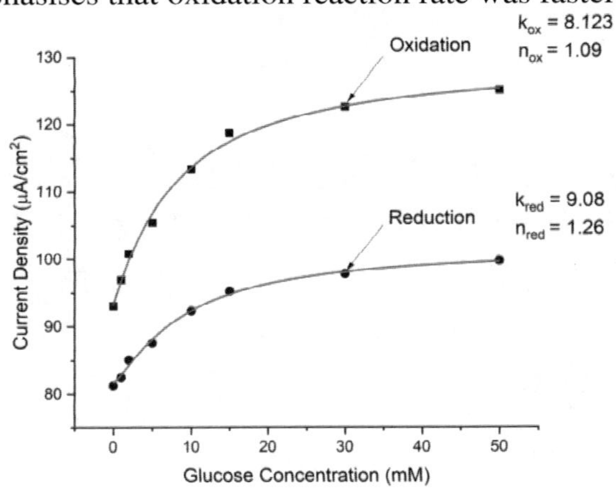

Figure 2. The peaks of current from voltammogram as a dependency on glucose concentration. Oxidation peak registered at +380 mV; Reduction peak registered at +150 mV.

IOP Conf. Series: Materials Science and Engineering 1239 (2022) 012006 doi:10.1088/1757-899X/1239/1/012006

The highest current density achieved at a glucose concentration of 30 mM. For that reason, other AuNP-modified graphite electrodes were tested at 9 pH and 30 mM glucose concentration.

Cyclic voltammograms were recorded at 750 μM of potassium ferricyanide, in pH 9 and 30 mM glucose concentration, using graphite electrode modified with different size gold nanoparticles, from 4 to 75 nm (Figure 3). Experimental data showed that the highest current density of 181 μA/cm^2 was achieved when particle size was 5 nm. And when particle size reaches 40 nm, current density decreases substantially and reaches plain graphite electrode.

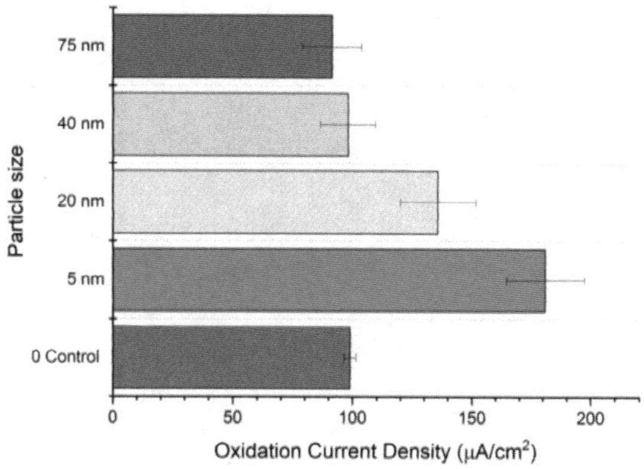

Figure 3. Oxidation peak current density dependency on particle size. Oxidation peak at +380 mV; Solution: PABS of pH 9, 0.75 mM of potassium ferricyanide, and 30 mM glucose.

The system with 5 nm AuNP's were evaluated by fitting the peak current was from voltammograms to Hill's equation (Equation 1) (Figure 4). The highest current density was achieved when AuNP size was 5 nm at a maximum peak current density of 183 μA/cm^2.

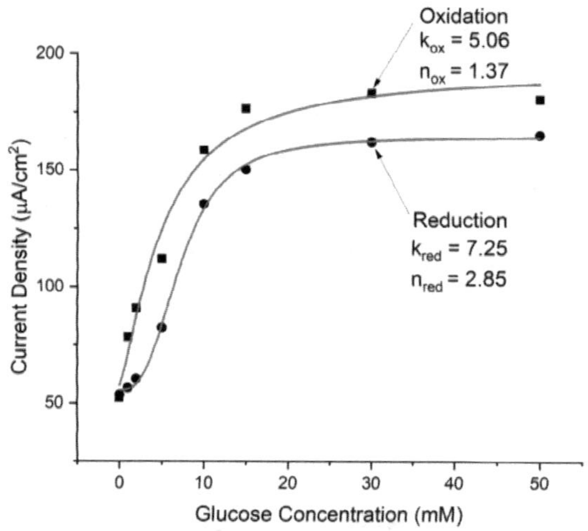

Figure 4. The peaks of current from voltammogram as a dependency on glucose concentration. When pH is 9; 0.75 mM of potassium ferricyanide. Oxidation peak at +380 mV; Reduction peak at +150 mV.

At the 0.15 V potential, the Hill coefficient n was higher than one (n=2.86), and at the 0.35 V potential, n was also higher than one (n=1.37), indicating that positive cooperativity was observed at oxidation and reduction peaks. It means if glucose concentration increases, then the efficiency of charge transfer by glucose molecules also increases. Half maximum concentration constant k (k_{ox}=5.06) was lower at oxidation potential than reduction (kred=7.25), emphasising that the reaction rate was faster at oxidation. The built system can still be improved by removing sulfur from gold particles, thus providing a larger surface area for the catalytic reaction.

Summary

A biofuel cell based on the gold nanoparticles concept was designed and evaluated. This research shows that the optimal medium pH for such biofuel cells would be 9. This could be because it can withstand a higher concentration of gluconolactone, which provides better efficiency in the system. After evaluating the optimal pH for the medium, it was used to research the impact of the gold nanoparticle size on the system performance. The highest current density (183 μA/cm^2) was observed when AuNPs size reached 5 nm, and glucose concentration was 30 mM. It increased linearly with decreasing size of the particles. Smaller size nanoparticles could be used to improve current density further. And according to Hill's equation, all reactions have positive cooperativity, however, reactions speed increased when smaller particles were used. As synthesis process uses sulfuric acid gold could have some sulfur residue on its surface, thus the reaction has less surface area to take place. This system could be further improved by the removal of sulfur from the gold surface. Our future research will be related to measuring power output and upgrading the current design of the system.

References

[1] R. A. Bullen, T. C. Arnot, J. B. Lakeman, and F. C. Walsh, "Biofuel cells and their development," Biosensors and Bioelectronics, vol. 21, no. 11. pp. 2015–2045, May 15, 2006, doi: 10.1016/j.bios.2006.01.030.

[2] I. Willner, Y. M. Yan, B. Willner, and R. Tel-Vered, "Integrated enzyme-based biofuel cells - A review," Fuel Cells, vol. 9, no. 1. pp. 7–24, Feb. 2009, doi: 10.1002/fuce.200800115.

[3] Q. Xu, F. Zhang, L. Xu, P. Leung, C. Yang, and H. Li, "The applications and prospect of fuel cells in medical field: A review," Renewable and Sustainable Energy Reviews, vol. 67. Elsevier Ltd, pp. 574–580, Jan. 01, 2017, doi: 10.1016/j.rser.2016.09.042.

[4] A. Nasar and R. Perveen, "Applications of enzymatic biofuel cells in bioelectronic devices – A review," International Journal of Hydrogen Energy, vol. 44, no. 29. Elsevier Ltd, pp. 15287–15312, Jun. 07, 2019, doi: 10.1016/j.ijhydene.2019.04.182.

[5] J. Kim, H. Jia, and P. Wang, "Challenges in biocatalysis for enzyme-based biofuel cells," Biotechnology Advances, vol. 24, no. 3. pp. 296–308, May 2006, doi: 10.1016/j.biotechadv.2005.11.006.

[6] X. Wang, W. Guo, Y. Hu, J. Wu, and H. Wei, Nanozymes: Next wave of artificial enzymes. 2016.

[7] L. Cao, P. Wang, L. Chen, Y. Wu, and J. Di, "A photoelectrochemical glucose sensor based on gold nanoparticles as a mimic enzyme of glucose oxidase," RSC Adv., vol. 9, no. 27, pp. 15307–15313, 2019, doi: 10.1039/c9ra02088h.

[8] L. Fan et al., "A Novel AuNP-Based Glucose Oxidase Mimic with Enhanced Activity and Selectivity Constructed by Molecular Imprinting and O2-Containing Nanoemulsion Embedding," Adv. Mater. Interfaces, vol. 5, no. 22, pp. 1–9, 2018, doi: 10.1002/admi.201801070.

[9] N. J. Lang, B. Liu, and J. Liu, "Characterisation of glucose oxidation by gold nanoparticles using nanoceria," *J. Colloid Interface Sci.*, vol. 428, pp. 78–83, 2014, doi: 10.1016/j.jcis.2014.04.025

[10] S. Biella, L. Prati, and M. Rossi, "Selective oxidation of D-glucose on gold catalyst," *J. Catal.*, vol. 206, no. 2, pp. 242–247, 2002, doi: 10.1006/jcat.2001.3497.

[11] J. N. Weiss, "The Hill equation revisited: uses and misuses," FASEB J., vol. 11, no. 11, pp. 835–841, 1997, doi: 10.1096/fasebj.11.11.9285481.

Evaluation of oxygen consumption at biologically active surfaces using scanning electrochemical microscopy

A Zinovicius[1], J Rozene[1,b], G Dirma[1], L Striska [1], S Stankaitis[1], and I Morkvenaite-Vilkonciene[1]

[1]Department of Mechatronics, Robotics, and Digital manufacturing, Vilnius Gediminas technical university, Basanaviciaus g. 28, Vilnius, Lithuania

Keywords: scanning electrochemical microscopy, oxygen, living cells, mathematical model, ultramicroelectrode.

Abstract. Scanning electrochemical microscopy (SECM) is a local electrochemical analysis technique, which performs scanning of the sample by ultramicroelectrode (UME) [1]. The result of measurement usually is current, which depends on the electrochemical properties of the surface of interest. The errors can be observed if UME geometry is not ideal [2,3]. This happens after UME polishing or breaking after contact with the sample surface. And the aim of this research was to evaluate oxygen consumption at biologically active surfaces using scanning electrochemical microscopy. That is why the oxygen consumption at the surface of interest with UME of different geometry was evaluated in this work. The model results were compared with experimental data. The computational experiment was performed with the SECM model using diffusion equations in the COMSOL program. This model could be further improved to measure diffusion rates with different probe geometries and coefficients.

1. Introduction

Scanning electrochemical microscopy (SECM) is a scanning probe microscopies technique that uses an ultramicroelectrode (UME) probe. It can be used to evaluate electrochemically/electrically active surfaces [1], biosensors [2], [3], and biofuel cells [4]. The current flows through a probe that can be positioned within 3 axis, and potentiostat registers electric current dependency on distance and displacement. The conductive part of the UME is usually relatively small, and the diameter can be 10 μm or even less. During the experiment, the electrode is positioned near the sample's surface (insulating, conductive or semiconductive) in an electrolyte. In the SECM, the current is carried by redox processes at tip and substrate and is controlled by electron transfer kinetics at the interfaces and mass transfer processes in electrolyte. SECM apparatus has four main parts which are needed for its operation. First, a low current bipotentiostat (≤pA) is used to precisely measure and control the current and potential at both a probe and a substrate. The next part which is required for SECM is the probe - ultramicroelectrode, which diameter is ≤25 μm. It has a conductive part in the middle that can be made from platinum, gold, graphite and surrounded by the glass insulator. The probe is the sensor of the

Content from this work may be used under the terms of the Creative Commons Attribution 3.0 licence. Any further distribution of this work must maintain attribution to the author(s) and the title of the work, journal citation and DOI.
Published under licence by IOP Publishing Ltd

IOP Conf. Series: Materials Science and Engineering 1239 (2022) 012007 doi:10.1088/1757-899X/1239/1/012007

system, since it measures the current, which is detected at the electroactive part of the tip. For the positioning of the probe, stepper motors or servo motors can be used. These motors make the movements in more coarse increments. Then precision is needed during measurements piezoelectric inchworm motors are used. The signal measured from the probe goes to the data acquisition system, which graphically represents the measured data. It also sends signals for the position controller, which moves the probe. When measuring sample, different modes can be used, two main modes are feedback and generation/collection. At generation collection mode, the generated species are collected at the electrode. Feedback mode can be positive or negative. Positive feedback is observed when the tip approaches conductive surface, and as a result current increases close to surface. If the sample is an insulator when the probe approaches it, there are no redox species, and current becomes smaller at the close distance from surface. This effect is called 'hindered diffusion', since the diffusion is limited due to small gap between surface and tip. Negative feedback mode can be used for the detection of probe geometry, which directly affects measurement results. For the investigation of biological surfaces, very comfortable way to detect probe geometry is to measure negative feedback approach curve without redox mediators in solution. In this case, current is oxygen reduction current, since oxygen is dissolved in solution and can be good 'redox mediator' in this case. In the negative feedback mode probe consumes oxygen. Current is detected at the tip according to the amount of oxygen which is reduced at the tip.
The aim of this research is to evaluate oxygen consumption at biologically active surfaces using scanning electrochemical microscopy using mathematical model.

2. Mathematical model

In experiments where feedback mode is used, the tip is immersed in a solution that contains electrolyte and redox species. In this case oxygen is redox mediator, which concentration directly proportional to current. The rectangular domain is used to represent a standard plane ultramicroelectrode. Diffusion of oxygen expressed by second Fick's law in the cylindrical coordinate system [5]:

$$\frac{\partial C}{\partial t} = D_{O_2}(\frac{\partial^2 C}{\partial r^2} + \frac{1 \partial C}{r \partial r} + \frac{\partial^2 C}{\partial z^2}$$
$$for\ 0 < t \leq T, 0 < z < d, 0 < r < r_{glass} \qquad (1)$$

Where T is duration of computational experiment, C is concentration of O_2 it is expressed as a function of time t and spatial coordinates r and z, r_{glass} radius of insulating part, r radius of electroactive part, D diffusion coefficient in water.

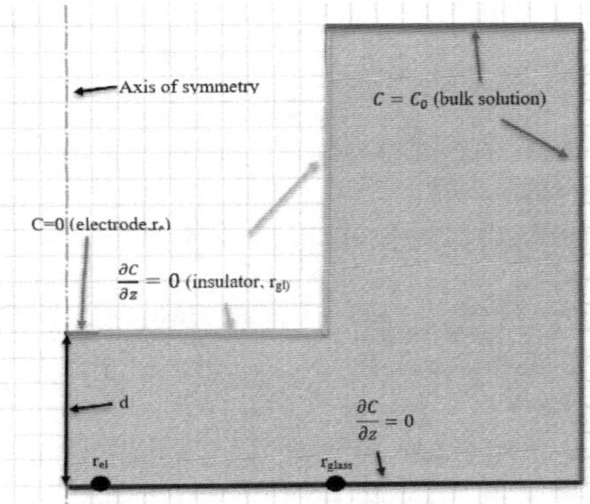

Figure 1 Geometry of simulation domain with the boundary conditions required for SECM.

At the start of the simulation, the solution is saturated with oxygen, and the concentration is set to 0.000253 mol/L. Boundaries which in if Fig. 1.1 is marked in blue color generate this concentration to simulate constant inflow of oxygen from solution. The surface of the electrode is set to consume oxygen, so the concentration on the electrode is set to zero:

$$\frac{\partial C}{\partial z} = 0 \tag{2}$$

There is a constant intake of oxygen from the outer side of the domain, which is marked in the blue line in Fig. 1.:

$$C = 0.000253 \frac{mol}{l} \tag{3}$$

The electrode is consuming oxygen, so its concentration is set to zero. It is marked with a red line in Fig. 1.:

$$C = 0 \tag{4}$$

The insulator is set to no flux boundary it means there is no flow of oxygen:

$$\frac{\partial C}{\partial z} = 0 \tag{5}$$

Current is calculated on the surface of an electrode according to this equation:

$$i = 2\pi F D_{O_2} \int_0^{r_{el}} \frac{\partial C}{\partial z} r dr \tag{6}$$

Where: n=4, it's equal to the number of electrodes exchanged. F is Faraday constant F=9.648*10^4C/mol and D_{O_2} is oxygen diffusion coefficient, which is 1.105*10^{-9} m^2/s.

The model described above is used to perform simulation of SECM negative feedback. The electrode tip is approaching an insulating surface that does not generate species detected by the electrode. Also, it could be changed to positive feedback mode if the isolating surface would be set to generate oxygen which then could be detected with the electrode.

3. Results

First experiments with real SECM were performed to measure negative feedback approach curves. For measuring disk-shaped UME-based probe with radius R_{gl}=60.5 μm, R_{el}=3.5 μm was used for experiments. The R_g value represents the ratio of insulating part radius with conducting part radius and was calculated as 17.3.

For the measurements with ultramicroelectrode, a three-electrode electrochemical cell setup was used. SECM probe was switched as a working electrode. A negative feedback approach curve was registered with a potential of -500 mV.

The results of the measurements is shown in Fig. 2. When the distance from isolating surface is the greatest, the current registered with the electrode is the largest. As the tip approaches the isolating surface, the oxygen around the electrode tip is consumed, and the bulk solution's oxygen cannot replace

IOP Conf. Series: Materials Science and Engineering 1239 (2022) 012007 doi:10.1088/1757-899X/1239/1/012007

it. The current depends on the amount of oxygen in the solution and the R_g ratio of the electrode used to measure.

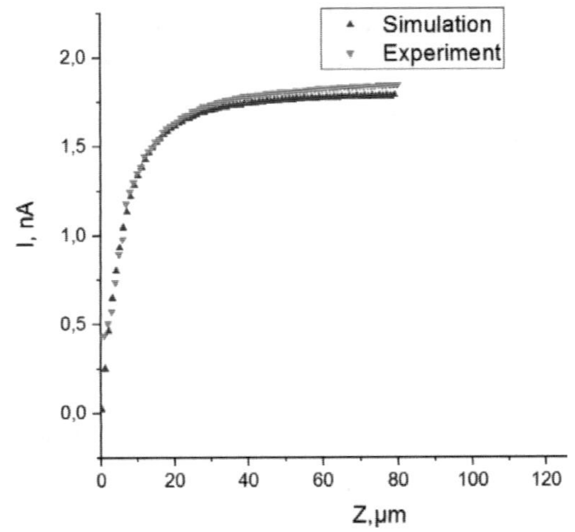

Figure 2. Negative feedback approach curve at -500 mV.

The same Fig. 1 has data from the mathematical model in the graph marked with blue triangles. The difference between experimental and simulation data can be visible then the distance is greater than 30 μm. Data from experiments has a higher steady-state current, but the approach curve is the same for both methods.

The error which can be seen in the graph comes from the imperfection of the experimental measurement. Real ultramicroelectrode geometry always has some flaws. Also, the oxygen concentration in the solution could be different from one in the simulation, and this error is what causes the difference between steady-state current curves.

Other simulation was performed to analyze how different probe geometries were performed in negative feedback mode than standard ultramicroelectrode. In the Fig. 3 approach curves, where the probe had a recessed geometry, are plotted. The recession of the electrode conductive part was increased by 2 μm. As the recession of the electrode increases, the steady-state current detected by the probe decreases. This effect happens because the glass insulator limiting oxygen diffusion is even more influencing the diffusion to the tip when the probe conductive part is recessed inside the glass insulator.

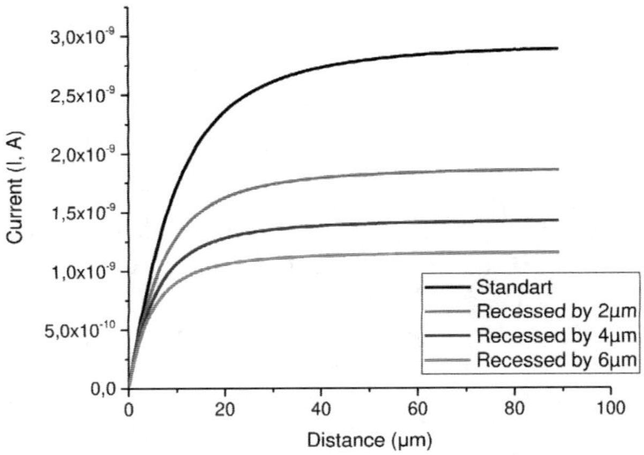

Figure 3. Negative feedback approach curves with different recessed probe geometry.

When the recession is increased from 2 to 6 μm, the steady-state current decreases from 1 to 2.75*10-9A , but then the distance from the surface to which the probe is approaching decreases, the current at the tip becomes the same for all geometries. At this point, diffusion is mainly limited by the diffusion coefficient and the consumption of oxygen by the probe. The following simulation was performed when the electroactive part of the probe was outwarded by increments of 2 μm. The resulted data can be seen in Fig. 4

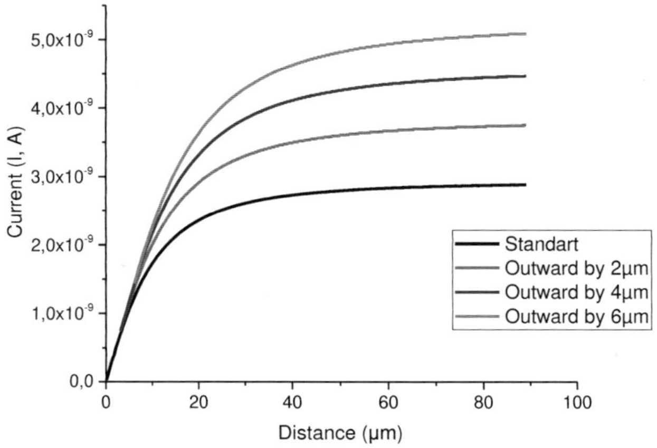

Figure 4. Negative feedback approach curves with outwarded geometry.

When the electrode geometry is outwarded, the steady-state current increases almost 2 times from 2.75 nA to 5 nA compared to the standard electrode. This happens because of the increase of the area of electroactive. So, in this case, two sides of the electroactive part can detect oxygen. Also, diffusion is less limited by the glass insulator since the whole conductive part of the probe is further away from the glass.

4. Conclusions

This article presents how the SECM probe can be modelled in a COMSOL environment. The model shows how the change in geometry can have a significant impact on the measurements. The outward probe geometry has a big influence on the steady state currents values detected at the probe because the electroactive surface area increases, and so does the detected current. When probe geometry has recessed conductive part, steady-state current decreases compared to standard ultramicroelectrode. The glass insulator limits the diffusion of oxygen and cannot reach the electroactive part. This model could be further improved to measure diffusion rates with different probe geometries and coefficients.

Acknowledgements

This project has received financial support from the Research Council of Lithuania (LMTLT), Nr. P-LLT-21-6, State Education Development Agency of Latvia, Ministry of Science and Technology (MOST) of Taiwan, Nr. MOST 110-2923-E-011-006-MY3.

References

[1] A. J. Bard, F.-R. F. Fan, and M. V Mirkin, "Scanning electrochemical microscopy," *Electroanal. Chem.*, vol. 18, pp. 243–373, 1994.

[2] I. Morkvenaite-Vilkonciene, A. Ramanaviciene, A. Kisieliute, V. Bucinskas, and A. Ramanavicius, "Scanning electrochemical microscopy in the development of enzymatic sensors

and immunosensors," *Biosens. Bioelectron.*, vol. 141, p. 111411, 2019.

[3] F. Ivanauskas, I. Morkvenaite-Vilkonciene, R. Astrauskas, and A. Ramanavicius, "Modelling of Scanning Electrochemical Microscopy at Redox Competition Mode Using Diffusion and Reaction Equations," *Electrochim. Acta*, vol. 222, pp. 347–354, Dec. 2016.

[4] A. Kisieliute *et al.*, "Towards microbial biofuel cells: Improvement of charge transfer by self-modification of microoganisms with conducting polymer – Polypyrrole," *Chem. Eng. J.*, vol. 356, pp. 1014–1021, Jan. 2019.

[5] R. Astrauskas, F. Ivanauskas, I. Morkvenaite-Vilkonciene, A. Ramanavicius, I. Morkvenaite-Vilkonciene, and A. Ramanavicius, "Mathematical Modelling of the Influence of Ultra-micro Electrode Geometry on Approach Curves Registered by Scanning Electrochemical Microscopy," *Electroanalysis*, vol. 31, no. 11, pp. 2214–2223, Jul. 2019.

IOP Conf. Series: Materials Science and Engineering 1239 (2022) 012008 doi:10.1088/1757-899X/1239/1/012008

Empirical analysis of piezoelectric stacks composed of plates with different parameters and excited with different frequencies

J Harazin[1], A Wróbel[1]

[1] Wydział Mechaniczny Technologiczny, Politechnika Śląska, Konarskiego 18A, 44-100 Gliwice, Poland

Abstract. Piezoelectric materials offer an ability to exchange energy between electrical and mechanical systems with fair ease, by using the simple and inverse piezoelectric effect. Piezoelectric transceivers, sensors, microphones, actuators, or active acoustic noise cancellation devices are some of many applications for piezoelectric materials. Due to the natural ability of the material to convert electricity and mechanical strain, solutions based on piezoelectric materials are often compact and allow applications on micro scales. One of the forms of piezoelectric actuation involves the use of piezoelectric stacks. The stack is composed of multiple piezoelectric plates layered by thin dielectric sheets, with electrodes attached along the sides. The main aim of piezoelectric stacks is the increase in maximum displacement by multiplying the number of piezoelectric plates that make up the stack. Stacks are composed of plates with the same material properties and the same dimensions. This study aims to investigate the idea of composing piezoelectric stacks of plates that have separate control circuits inducing different carrier frequencies or plates with differing properties and dimensions, in search for new applications for piezoelectric stacks. The main point of interest is the investigation of the ability to use piezoelectric stacks to generate complex vibration spectrums composed of multiple frequencies, resulting from the use of different piezoelectric plates in the stack or different carrier frequencies that stimulate each plate. To achieve this, a stack composed of two piezoelectric plates, each controlled by its own circuit, will be measured by a laser vibrometer, to check the complexity of the output vibration pattern.

1. Introduction

Piezoelectric materials contain a specific crystalline structure that gives them an ability to exchange energy between electrical and mechanical systems. The relations between mechanical strain and electric charge are defined by simple and inverse piezoelectric effects [1,2]. Currently, the most dominant group of piezoelectric materials are piezoceramics and single piezo crystals which are tested and developed since 1950s and 1960s, respectively. They come in many different available alloy compositions (notably containing elements such as lithium, titanium, niobium, barium, zinc, lead) depending on the desired set of material properties [2,3,4]. The R&D has also been put into new, alternative types of piezoelectric materials based on polymers and alloys that do not contain lead. Materials containing both existing and new alloys are also formed into composites with other ceramics or polymers to further extend the possible array of use cases and adhere to new norms [5,6,7]. The most common forms, in which piezo materials are seen in modern industry are differently sized and shaped plates, foils or sheets and assemblies or composites made from the above. Due to natural ability of converting electricity and mechanical strain, solutions utilising piezoelectric materials are often compact and allow applications on micro scales. Piezoelectric transceivers,

Content from this work may be used under the terms of the Creative Commons Attribution 3.0 licence. Any further distribution of this work must maintain attribution to the author(s) and the title of the work, journal citation and DOI.
Published under licence by IOP Publishing Ltd

sensors, microphones, actuators, or active acoustic noise cancellation devices are some of many applications for piezoelectric materials [7,8,9,10]. The main disadvantage of piezoelectric materials is their small mechanical displacement caused by the inverse piezoelectric effect in relation to their size. Homogenous piezoelectric crystals made from continuous crystalline structure tend to generate the best displacements in relation to their size and applied voltages but are expensive to produce and are not always an ideal solution. An alternative method involves combining piezoelectric plates into piezoelectric stacks. An exemplary piezoelectric stack is usually composed of multiple piezoelectric plates layered by thin dielectric sheets, with electrodes attached along the sides (Fig.1a). This helps in achieving greater displacements, while sacrificing more space but at lower costs. Alternatively, coupled with some simple mechanisms (like a lever) similar displacements can be achieved with even more compact solutions by sacrificing part of generated force. That type of piezoelectric actuator is obtained by encasing the stack in a specially designed case, creating an amplified piezoelectric actuator (Fig.2b). So far, the main goal in using actuators made up of several piezoelectric plates is the increase in generated displacements. Piezoelectric stacks are made from plates of near identical properties and dimensions to avoid unwanted deviations in their response to stimulating current and to maximise their efficiency. The focus is also put into mathematical modelling of such piezoelectric actuators to further optimise their working parameters and increase their efficiency [13,14].

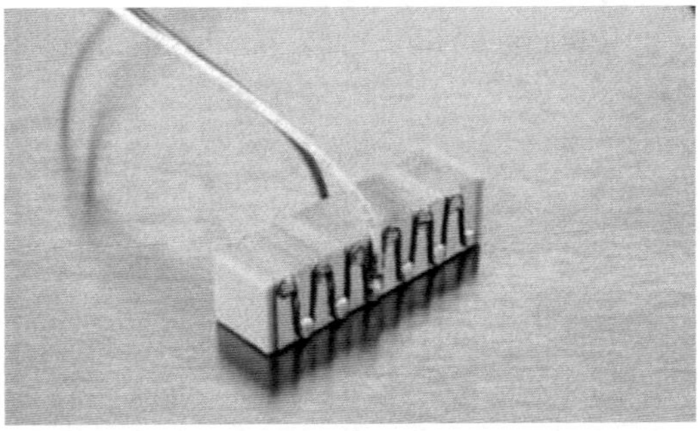

Figure 1a. An example of piezoelectric stack composed of several rectangular piezoelectric plates, with electrodes attached on the sides [12]

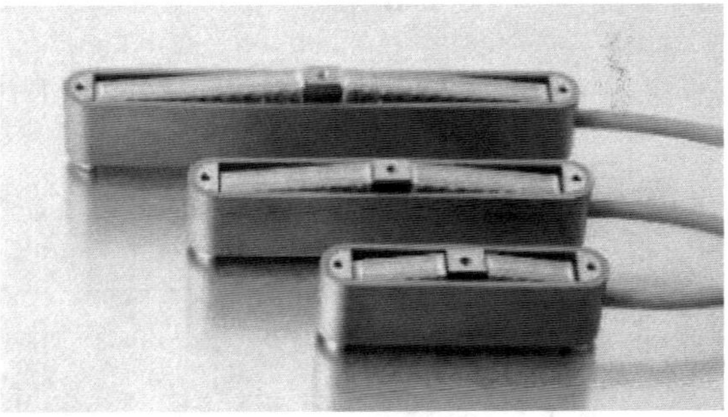

Figure 1b. An example of an amplified piezoelectric actuator, the stack is placed in a specially designed case, that multiplies the displacement generated by the piezoelectric [12]

As the highlight of this article, a focus is being put on the possibility of using piezoelectric materials in active vibration cancellation techniques. As part of our research, we are investigating the possibility of using piezoelectric stacks to dampen vibrations with complex spectrum of frequencies, by treating plates composing the stack as separate stages which are responsible for damping parts of the spectrum. This solution could effectively combine the benefits of greater displacements, already offered by existing solutions with additional applications in active vibration damping systems. This article in particular is concentrated around the study of piezoelectric stacks in which each plate composing the stack is coupled with a separate control circuit, enabling it to be stimulated independently of others. The main aim of this study is to determine the possibility of presented thesis and to find any important factors and possible obstacles that must be considered in designing systems based around this idea.

2. Methods

2.1. Samples and equipment

The idea for the study included measurements of actuation frequency for piezoelectric plates, paired together in small "two stage" stacks with the use of measuring equipment already available in the laboratories of the faculty of Mechanical Engineering. Sample piezoelectric plates for the study,

IOP Conf. Series: Materials Science and Engineering 1239 (2022) 012008 doi:10.1088/1757-899X/1239/1/012008

kindly provided by Ferroperm Piezoceramics, included plates of different dimensions, made from material called Pz26:

- 18x18x8 mm (1 plate)
- 30x30x10 mm (2 plates)

All the information about the properties of piezoelectric material and the actuation response for each of the abovementioned plates were obtained from the Ferroperm Piezoceramics site [15].

A waveform generator from Tabor Electronics was used to generate actuation current for both tested piezoelectric plates. A model named "WW5064" was used in the study. It was capable of generating 4 different waveforms simultaneously, with peak amplitude of 5 V and maximum frequency of 20 Mhz. To generate waveforms with different frequencies, an "arbitrary" mode was used, in which all frequencies were generated by a single internal clock but could be modulated by changing the density of sections in the waveform (1024 sections were standard, less would mean higher frequency, more – lower frequency). A detailed specification of the generator can be obtained from [16].

Piezoelectric plates need substantially higher voltages for generating a noticeable displacement, so an amplifier was required to amplify voltages generated from the generator. For that purpose, two identical amplifiers from Smart Material, already available at the laboratory were used. The model number of the amplifier was "HVA 1500/50" and it could multiply the input voltage 200-fold. The initial thought for the study assumed measuring the actuation of each plate in their respective resonant frequency to maximise displacements obtained from each plate in the stack but unfortunately due to physical limitations of the amplifier (maximum effective frequency range was ending at just 10 kHz), that idea had to be postponed for further studies. The detailed specification of the amplifiers can be found at [17].

The measurement of stack displacement was planned to be conducted through either the use of strain gauges coupled with a measuring amplifier or using a laser vibrometer. After initial tests, the laser vibrometer turned out to be better suited for the study because of much higher resolution and more flexible application. The laser vibrometer used for the study was made by Polytec. Its model number was "OFV-574" for the laser emitter and "OFV-2570" for the laser controller. The laser was capable of registering displacements down to μm with a maximum frequency of 10 MHz. The full specification of laser vibrometer can be seen at [18].

The output terminal from vibrometer was forwarded to an oscilloscope from Rigol with a model number DS1064B. The data was then obtained by doing a series of measurements, saved to a CSV format and exported to the computer via a USB stick. Full specification of the oscilloscope is presented in [19].

2.2. Experiment procedure

A small vice was used to fix the measured piezoelectric stack in place. Places of direct contact between the surfaces of the vice and the piezoelectric stack were lined with two layers of rubber tape to electrically isolate the surfaces and enable more free propagation of vibrations generated by the piezo stack. The stack was composed of two plates separated by several (3-4) sheets of very thin, microscope glass coverslips, glued together with cyanoacrylate adhesive which formed a makeshift composite material with thickness of around 1-2 mm. That way, a proper electrical isolation between plates was assured together with making enough space to fit the electrodes inside the coupling. Plates and glass sheets were slightly compressed and held together with clamps, until the adhesive settles and connects all surfaces evenly. For the piezo electrodes, a very thin and small plates of tin were cut out from tin sheet. The plates were soldered to the silver coating on the sides of piezoelectric surfaces. The plates were soldered on the opposite sides of the connected plates to save space and avoid any sparks jumping between the contacts and the other plate in the stack. To be certain that both plates are fully isolated, a small snippet of rubber tape was also fit in the remaining space between the soldered contact and the other plate in the stack. A laser was pointed at one face of the stack as precisely as possible to ensure that the reflected beam was pointing back at the vibrometer receiver. A special reflective strip was glued on the surface of the plate at the place where the laser was pointing to ensure good reflection. A draft of the setup can be seen in (Fig.2).

Two tests were prepared to measure the behaviour of piezoelectric stack composed of different plates. All measurements were done for the actuation voltage set to 2 V on the generator (400 V on the output of the amplifier). The tests were conducted for a simple sine wave and a simple triangle wave. The first test included two identical piezoelectric plates (30x30x10) and the second one included

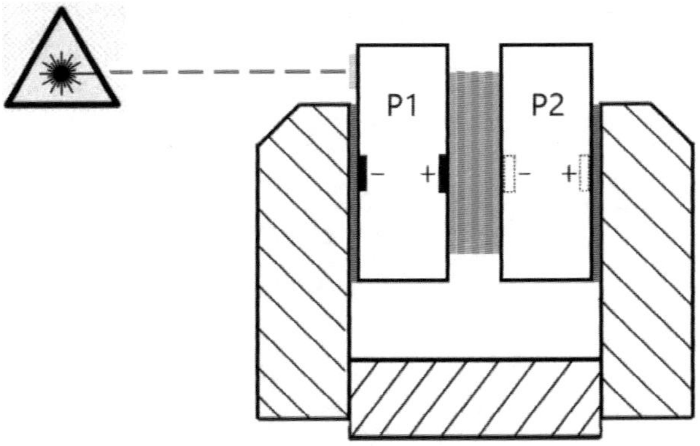

Figure 2. A simplified cross section of the test setup. P1 and P2 are symbolic representations of each piezoelectric plate in the stack. The laser warning sign represents the vibrometer as the source of the laser beam. The reflective patch was marked in green; the rubber tape was marked in red and the glue/glass composite was marked in orange/blue, respectively. This representation is not representing true sizes of each component.

a 30x30x10 plate connected with a 18x18x8 one. In both tests, the 30x30x10 plate was the forward-facing plate, at which the laser was pointing at. Three arbitrary frequencies of 5,8 and 10 kHz were chosen. The tests were conducted in the following combinations:

- 10 kHz on the first plate – 10 kHz on the second plate
- 10 kHz on the first plate – 8 kHz on the second plate
- 10 kHz on the first plate – 5 kHz on the second plate
- 8 kHz on the first plate – 10 kHz on the second plate
- 5 kHz on the first plate – 10 kHz on the second plate

Each series of tests were conducted for the sine wave and the triangle wave. Each test was recorded with an oscilloscope reading to the CSV file, which was later refined and combined with other readings in Excel.

3. Results

Due to rather extensive scope of comparisons and aggregations of the acquired data, we focused on showcasing only the most prominent differences and irregularities in the dataset. Full results of the tests were saved and are being stored on the Mendeley data site. The full mention to results presented in this article was placed in the data availability section.

The biggest displacement for the stack composed of the same plates was 0,035 μm in the 10_10 kHz configuration whereas for the stack of different plates and the same frequency configuration, the maximum obtained displacement was 0,021 μm. The static displacement for both 30x30x10 and 18x18x8 plates at 400 V calculated from [15] was 0,131 μm for a freestanding plate. The sum of both plates' displacement should be equal to 0,262 μm irrelevant of the plates used in the setup according to the data from the calculator with an error of up to 20% for a single plate.

As for the difference between actuation with a sine wave and a triangle wave, the triangle wave has seen overall worse performance in comparison to sine wave in terms of displacement generated on both configurations of the plates. The difference ranged between 13% - 20% less displacement for the triangle wave with an average of 18,9%.

The difference between displacement generated by the first plate and the second plate has shown that registered displacements of the second plate (P2 on Fig. 2) were lower by 62 – 66% for the 10kHz and

8 kHz frequency with the same plates in the stack. For different plates and a frequency of 10kHz and 8 kHz the displacements of plate P2 were respectively lower by 84 – 86%. There was an anomaly however, for the registered displacements of plates for the frequency of 5 kHz. For the setup with the same plates in the stack, the registered displacement of the second plate P2 was 74% bigger than the displacement from plate P1. For the setup with different plates P2 was vibrating with only 50% smaller amplitude than P1 instead of expected 84 – 86%. The summary of this data has been shown in Table 1.

Table 1. The summary of differences in displacements between P2 and P1

Difference between displacement of 2nd and 1st plate						
30x30x10 + 30x30x10			30x30x10 + 18x18x8			
10kHz	8kHz	5kHz	10kHz	8kHz	5kHz	
-0,019	-0,009	0,003	-0,021	-0,025	-0,004	**[μm]**
-66,9436	-62,2374	74,972	-84,4015	-86,2277	-49,964	**[%]**

Graphs showcasing FFT transforms of each set of displacements and their comparisons are available online on the *Mendeley* data storing site.

4. Discussion

The sum of displacements for both plates in the stack exceeded the predicted displacement calculated from the calculator provided in [15] by 34% which raises concern as to the accuracy of either the measured results or the prediction made from the calculator. However, the calculator was designed to measure static displacements and only for single plates so the deviations are to be expected but it should be noted that a better and more accurate way to predict and measure the displacements must be created.

The aperture that was available for this study did not have the ability to generate complex waves with multiple component frequencies which could be applied to measure the displacement of just a single plate. A comparison between using multiple plates with simple waveforms against using a single plate with a complex waveform would provide a good perspective on the meaningfulness of the conducted study and should be conducted in near future.

The causes of the anomaly in the registered displacements of plate P1 and P2 in the 10_5kHz and 5_10kHz configurations must be investigated further. For now, either an error in the measuring process or an interference resulting from a complex structure of the examined stack and harmonic frequencies are the hypothetical causes of this anomaly. The structure of the test stand has to also be improved because in the present state it could have been the source of many inconsistencies.

5. Results

The main goal of the study (investigating the possibility of using differently modulated and different plates in a single stack) was achieved along with pinpointing at least part of existing issues and obstacles that are connected to the proposed idea. This study showed that the conception is still immature and needs a lot more research to prove its value and meaningfulness over other existing methods. It is necessary to conduct experiments with plates in their resonant frequency to maximise the potential of this conception and refine methods of gathering and comparing data to avoid uncertainties and errors. It is also important to compare the results of this proposed method with alternative and existing solutions to determine its usefulness. The issues raised and comments highlighted will be the focus for the following studies, to narrow down the uncertainty that is still associated with this idea.

This study has shown that the idea itself has a possibility to increase the flexibility of systems using piezoelectric stacks as form of actuation and can enable the application of piezo stacks in systems that require precise actuation with broad spectrum of frequencies.

IOP Conf. Series: Materials Science and Engineering 1239 (2022) 012008 doi:10.1088/1757-899X/1239/1/012008

6. Data availability

The entirety of data used for this study, including raw input information and aggregated results was stored on the Mendeley Data site under this address: http://dx.doi.org/10.17632/yfr68bw58c.1

7. References

[1] Jiashi Y 2018 An introduction to the theory of piezoelectricity *Advances in Mechanics and Mathematics*, ISBN 978-3-030-03137-4

[2] Soluch W 1980 Wstęp do piezoelektryki *Wydawnictwo Komunikacji i Łączności* (Poland), ISBN 83-206-0041-3.

[3] Guo Q, Li F, Xia F, Wang P, Gao X, Hao H, Liu H, Sun H, Zhang S 2020 Piezoelectric ceramics with high piezoelectricity and broad temperature usage range, *Journal of Materiomics*, ISSN 2352-8478, doi: https://doi.org/10.1016/j.jmat.2020.11.012

[4] Berlincourt D 1981 Piezoelectric ceramics: Characteristics and applications *The Journal of the Acoustical Society of America* vol. 70 pp. 1586-1595, doi: https://doi.org/10.1121/1.387224

[5] Huige W, Hui W, Yijie X, Dapeng C, Yapeng S, Mengyao D, Chuntai L, Tao D, Jiaoxia Z, Yong M, Ning W, Zicheng W, Ye S, Renbo W, Zhanhu G 2018 An overview of lead-free piezoelectric materials and devices (*J. Mater. Chem. C The Royal Society of Chemistry* vol. 6 issue 46) pp. 12446-12467, doi: 10.1039/C8TC04515A

[6] Shrout T R, Zhang S J 2007 Lead-free piezoelectric ceramics: Alternatives for PZT? *J Electroceram* vol. 19 pp. 113–126, doi: https://doi.org/10.1007/s10832-007-9047-0

[7] Maity K, Mandal D 2021 Chapter Seven - Piezoelectric polymers and composites for multifunctional materials (*Advanced Lightweight Multifunctional Materials by Costa P, Costa C M, Lanceros-Mendez S Woodhead Publishing in Materials*) pp. 239-282, doi: https://doi.org/10.1016/B978-0-12-818501-8.00001-9

[8] Duan W H, Wang Q, Quek S T 2010 Applications of Piezoelectric Materials in Structural Health Monitoring and Repair: Selected Research Examples *Materials* vol.3 pp. 5169-519, doi: https://doi.org/10.3390/ma3125169

[9] Li H, Tian C and Deng Z D 2014 Energy harvesting from low frequency applications using piezoelectric materials *Applied Physics Reviews* vol. 1, doi: https://doi.org/10.1063/1.4900845

[10] Goldfarb M, Celanovic N 1997 Modeling piezoelectric stack actuators for control of micromanipulation *IEEE Control Systems Magazine* vol. 17 no. 3 pp. 69-79, doi: 10.1109/37.588158.

[11] Zhou Q, Lam K H, Zheng H, Qiu W, Shung K K 2014 Piezoelectric single crystal ultrasonic transducers for biomedical applications *Progress in Materials Science* vol. 66 pp. 87-111, ISSN 0079-6425, doi: https://doi.org/10.1016/j.pmatsci.2014.06.001.

[12] Information on http://www.noliac.com/products/

[13] Adriaens H J M T S, De Koning W L and Banning R 2000 Modeling piezoelectric actuators *IEEE/ASME Transactions on Mechatronics* vol. 5 no. 4 pp. 331-341, doi: 10.1109/3516.891044.

[14] Rodriguez-Fortun J M, Orus J, Buil F, Castellanos J A 2010 General Bond Graph model for piezoelectric actuators and methodology for experimental identification *Mechatronics* vol. 20 Issue 2 pp. 303-314, ISSN 0957-4158, doi: https://doi.org/10.1016/j.mechatronics.2010.01.004.

[15] Information on https://www.meggittferroperm.com/resources/piezo-calculator/

[16] Information on https://www.taborelec.com/ww5064#Highlights

[17] Information on https://www.smart-material.com/HVAsV2.html

[18] Information on https://www.polytec.com/eu

[19] Information on https://www.batronix.com/shop/oscilloscopes/Rigol-DS1064B.htmls

IOP Conf. Series: Materials Science and Engineering 1239 (2022) 012009 doi:10.1088/1757-899X/1239/1/012009

Surface laser processing of maraging steel parts manufactured by selective laser melting: effect on pass geometry and hardness

Kęstutis Bučelis[1], Jelena Škamat[2], Olegas Černašėjus[1]

[1]Department of Mechanics and Materials Engineering of Vilnius Gediminas Technical University, J. Basanavičius str. 28, LT-03224, Vilnius, Lithuania
[2]Laboratory of Composite Materials of Vilnius Gediminas Technical University Linkmenų Str. 28, LT-08217, Vilnius, Lithuania

kestutis.bucelis@vilniustech.lt

Abstract. Due to the possibility to produce the parts with complex internal and external geometries, selective laser melting (SLM) process attracts growing interest in various fields of engineering segments such as aircraft, aerospace, biomedical, automotive, marine industries and tooling. Maraging steels, having excellent weldability and high resistance to thermal fatigue due to the lack of carbon, has showed good suitability for SLM [1]. However, owing to the limited hardness and wear resistance, maraging steels has limited application at the harsh wear conditions [4]. In this study, the possibility to improve the surface characteristics of DIN 1.2709 steel SLM parts by application of laser alloying technology is evaluated. The surface of SLM part was laser processed at various laser spot diameters and varying laser scanning speeds from 500 to 1500 mm·min^{-1}, with and without preposition of alloying element. The power density was provided in the range from ~0.8·10^3 W·cm^{-2} to ~51·10^3 W·cm^{-2} and heat input – from 4 to 12 J·mm^{-1}. The effect of laser processing parameters and presence of alloying element on the geometry of obtained processed passes and hardness of surface was evaluated. It was determined, that the application of CO_2 continuous laser at the parameters of 1 kW laser power, 0.5 mm laser beam spot diameter and laser scanning speed in the range between 500 mm·min^{-1} and 1250 mm·min^{-1} allows obtaining laser pool of acceptable geometry and sizes directly on as-manufactured SLM part surface without any pre-processing. The increasing scanning speed to 1500 mm·min^{-1} or spot size to 2.0 and 3.0 mm results in too small pool depth and unstable pool geometry. The laser processing with preposition of alloying element layer provided surface alloying effect of the maraging steel SLM part. The hardness of processed surface areas ranged between ~600 HV0.2 at the lowest scanning speed and ~1770 HV0.2 at the highest speed, what is from 18% to ~3.5 times higher, as compared with maximum hardness of 1.2709 maraging steel after aging (~58 HRC or ~510 HV).

1. Introduction

Selective Laser Melting (SLM) is process utilizing high energy laser beam to selectively fuse powdered material and produce 3D parts in such a way [1]. It is one of the main processes of Additive Manufacturing (AM), which was developed for metal parts production. Fusing of powders and build-up of volume is performed "layer by layer". Using digital description of each cross-section, laser scans the pre-posited powders and melts them or leave un-melted. The thickness of each layer corresponds to the

Content from this work may be used under the terms of the Creative Commons Attribution 3.0 licence. Any further distribution of this work must maintain attribution to the author(s) and the title of the work, journal citation and DOI.
Published under licence by IOP Publishing Ltd

melt pool depth, which laser produces by one step. Such a technique allows producing parts with complex internal and external geometry and, therefore, it attracts growing interest in various fields of engineering, for which the components of complex geometry are typical (aircraft, aerospace, tooling) [2].

The heating and cooling of material during SLM process is very rapid; moreover, material experiences numerous reheating cycles when laser beam melts neighbouring passes or next layers. Therefore, the list of alloys suitable for SLM process is limited and carbon-free steels are usually used, which are not sensitive to thermal cracking. 18-percent nickel maraging steel is suitable one.
Maraging steels possess very high yield strength and high fracture toughness. However, the hardening is due to precipitation of intermetallics, which are not extremely hard. As a result, maraging steels have moderate hardness (~55–58 HRC max.) and, often, insufficient wear resistance. Laser surface alloying of metals is known as effective technique in term of improving the metal parts surface performance, including their hardness and wear resistance. Laser boronising of steels is one of the possible processes, in which the surface layer of the steel part being laser-melted and mixed together with pre-posited boron-containing layer. Boron solubility in iron is very low and, according to Fe-B phase diagram [3], iron forms with boron two types of borides – FeB (16.25 wt.%) and Fe_2B (8.48 wt.%), presence of which provides high hardness and wear resistance. Thermochemical boronising of metals has been extensively studied for various types of alloys, including maraging steels [4]. Laser boronising of steels and other alloys has not been studied so widely. To the best of authors' knowledge, at the moment, there are no data on the laser surface alloying of AM-manufactured parts produced from maraging steels.
The present work is aimed at primary evaluation of the effect of laser boronising process and its parameters on the geometry and hardness of the processed surface of 18Ni-300 maraging steel part manufactured by SLM process.

2. Experimental details

2.1. Material and method

DIN 1.2709 maraging steel, also known as 18Ni-300 maraging steel, was investigated. This is one of the most widely used and commercially available maraging steel grade. The samples for testing were produced by SLM process using DIN 1.2709 maraging steel powder (7–30 µm). The chemical composition of powder (according to powder manufacturer) is presented in Table 1. Prismatic samples with the dimensions 15×15×10(h) mm were built-up using 'Concept Laser M3' SLM equipment. More detailed description of the procedure is presented in [5].

Table 1. Chemical composition of DIN 1.2079 steel powder [% wt]

Fe	C	Si	Mn	Ni	Mo	Ti	Co	Al
balance	0.03	< 0,1	< 0,1	17-19	4.8	< 0,8	8.5-9.5	< 0.1

For laser processing of the SLM samples surfaces, continuous CO_2 laser (wave length – 1064 nm) was used and the processing parameters were as follows: laser power – 1 kW; laser spot diameter – 0.5, 2.0, 3.0 and 4.0 mm; processing speed – 500, 750, 1000, 1250 and 1500 mm·min^{-1}. Laser processing experiments were performed on the side-on surfaces of the built-up SLM samples without any mechanical or other pre-processing. As was earlier determined in our studies and showed by other researchers numerous times, the morphology and roughness parameters of top and side-on surfaces of as-manufactured SLM parts differ significantly. This is due to peculiarities of SLM process. The exact morphology and roughness are determined by a number of factors, such as the size of the powder particles, parameters of the laser fuse process, trajectory of the laser scanning. In the present research, the roughness of the surface to be processed was ~Ra=9.2 µm, its morphology is showed in Fig. 1. Before the laser processing, amorphous boron was deposited on the surface to be processed. As the reference, the specimens laser-processed at the same parameters without preposition of amorphous boron were used.

Fig. 1. Scanning electron microscope image of the side-on surface of as-manufactured SLM part

In order to evaluate the influence of the laser process parameters and presence of boron on the geometry of the fused and alloyed passes, the processed specimens, both re-melted with and without boron in various speeds, were cut into transverse sections by the water jet cutter, polished with various grades of polishing SiC papers followed with the diamond suspensions of grade from 9 to 1 µm and 0.2 µm fumed silica suspension. After cleaning the prepared cross-sections with isopropyl alcohol, the geometry of the obtained passes was studied under the optical microscope Nikon Eclipse MA200 and the microhardness of processed surface zones was measured using Zwick Roell ZHµ microhardness tester and 200 g load for 10 s.

3. Results

3.1. Laser pool geometry

For the surface alloying process, laser is operated at the melting mode, which is provided by the appropriate processing parameters, first of all – laser power, spot size, operating speed. Laser pool geometry varies with variation of these parameters – typically, width and depth of the pool increase with the increase in laser power and decrease in operating speed [6]. However, what amount of laser light will be absorbed and transformed into a heat depends on a number of other factors, such as physical properties of part material, surface chemical composition and topography, size of the processed part, focal distance, etc. Therefore, primary experimental study is often required to know exact pool geometry and to choose appropriate parameters (including hatch spacing) providing even modified layer of required thickness. SLM parts is typically quite rough with surface containing some residual pores. According to the results already showed by other researches [7], laser processing of SLM products could significantly diminish their roughness and seal pores. In this point of view, further development of SLM parts surface laser processing, which would allow combining polishing and alloying effects, seems to be promising. For this purpose, laser processing experiments in present research were performed without any surface pretreatment. 1 kW laser power and 0.5 mm, 2.0 mm, 3.0 mm and 4.0 mm spot sizes were applied. Taking into account the approximate laser efficiency of ~10 % the laser beam power was ~100 W. Using a laser power P and a laser beam spot diameter d, the power density (P/d, W·m^{-2}) can be calculated. The approximate power density values applied in this work are given in Table 2. Laser beam power P and operating speed v determine the heat input (P/v, J·m^{-1}), which also influences the pool geometry. Heat input applied in this work at various laser operating speed is given in Table 3.

Table 2. Applied power density at different laser spot size

Parameter	Value			
Laser spot diameter, cm	0.5	2.0	3.0	4.0
Laser spot area, cm^2	0.0019625	0.0314	0.07065	0.1256
Laser beam power, W	100	100	100	100
Applied power density, W·cm^{-2}	50955	3184	1415	796

Table 3. Heat input at different laser scanning speed

Parameter	Value				
Laser scanning speed, mm·min^{-1}	500	750	1000	1250	1500
Heat input, J·mm^{-1}	12	8	6	4.8	4

The microscopical analysis of the processed samples showed that more or less acceptable pool geometry was obtained at 0.5 mm laser spot diameter, corresponding to a ~5·10^4 W·cm^{-2} power density. The optical images of the re-melted zone cross sections of specimens laser-processed with and without boron addition along with the intervals for obtained pool widths and depths are given in Table 4. For the samples, processed without boron addition, pool width varied between ~560 and ~960 μm, pool depth – between ~75 and 285 μm. Noticeable, the variation of depth and width was determined within the pass. Most likely, it can be related with uneven rough primary morphology of the surface, resulted in the variation of focal distance. In the literature, the ~10^3–10^5 W·cm^{-2} power density interval is indicated as typical for metal melting mode [8].

Table 4. Laser-processed SLM steel part: geometry of passes, obtained with and without boron

Laser spot size/ scanning speed	Without boron		With boron	
	Optical micrograph of cross-sectional pass view	width / depth, (μm)	Optical micrograph of cross-sectional pass view	width / depth, (μm)
0.5 mm / 500 mm·min^{-1}	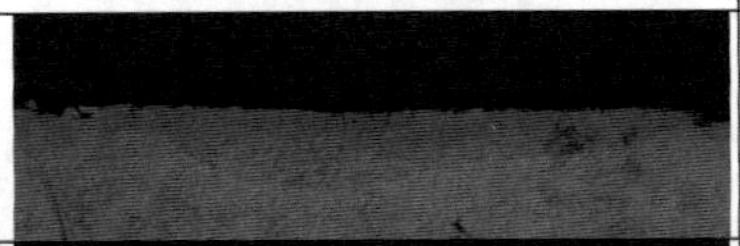	956-919 / 284-147		912-876 / 154-149
0.5 mm / 750 mm·min^{-1}		863-730 / 250-104	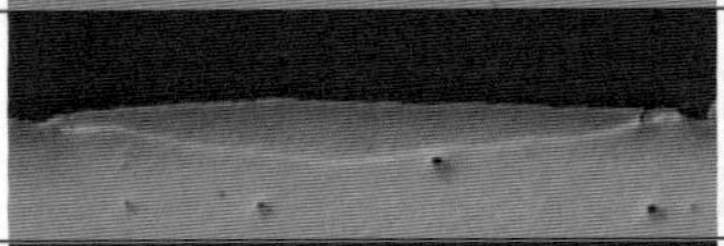	905-898 / 112-110
0.5 mm / 1000 mm·min^{-1}		835-658 / 179-81	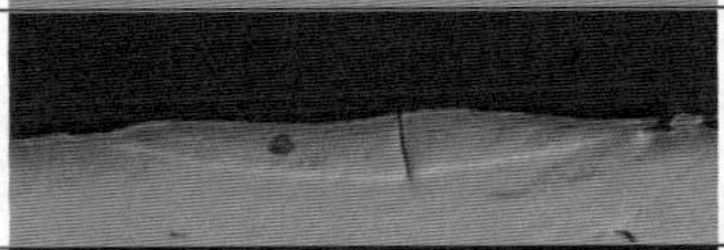	886-842 / 112-110
0.5 mm / 1250 mm·min^{-1}		758-568 / 175-119	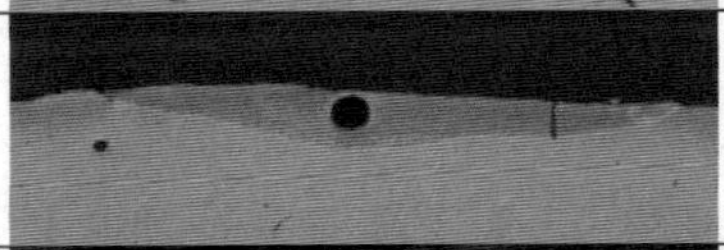	876-865 / 116-108
0.5 mm / 1500 mm·min^{-1}		738-559 / 107-75		853-741 / 57-54

The melting of surface occurred also applying 2.0 and 3.0 mm spot diameters, which provided power density in denoted interval; however, these applied regimes allowed obtaining neither stable pool width nor thickness. With 4.0 mm spot diameter, only heating of surface occurred.

With the pre-position of boron prior to the laser processing, the chemistry and morphology of the surface were changed, what could influence the absorbance of the laser energy. However, obtained laser pools did not differ visibly from those obtained without boron. At the same, the pores and transvers cracks were observed in all the samples processed with boron at scanning speeds 750 mm/min and over (heat input ≤ 8 J·mm^{-1}).

3.2 Hardness

The microhardness of as-manufactured SLM sample was ~360 HV0.2. The hardness of areas laser-processed without boron ranged between ~360 and ~540 HV0.2. The slight variation of the hardness may be related with the some aging effect due to samples heating during laser processing. Samples, laser-processed with addition of boron, showed higher hardness – from ~600 HV0.2 at the lowest scanning speed to ~1770 HV0.2 at the highest speed, what allows one to state that the alloying effect was reached. The hardness of FeB and Fe_2B borides is in the 17000–22000 MPa and 14000–18000 MPa ranges, respectively; here, lower limit corresponds to a hardness of pure borides, upper – borides alloyed with Ti, V, Cr, Mo; Cu, Ni and Al lower the hardness of iron borides by ~2000–3000 MPa [9]. Maraging steel, investigated in the present paper, contains both elements increasing the hardness of borides (Ti, Mo) and decreasing it (Ni, Al). Therefore, it is not able to identify the phases of the obtained layers based on their hardness – phase analysis is required. At the same time, pronounced trend of hardness increase with scanning speed, along with the appearance of transverse cracks (problem, typical for boride layers), shows that heat input may influence the content of boron. In this point of view, two possible mechanisms required further investigation: 1) the higher heat input causes partial burning out of boron (sublimation temperature ~2550°C); 2) the reduced size of steel melting pool at higher scanning speed results in higher concentration of boron after mixing.

Table 5. Microhardness HV0.2 of the processes surface zones

	0.5 mm 500 mm·min^{-1}	0.5 mm 750 mm·min^{-1}	0.5 mm 1000 mm·min^{-1}	0.5 mm 1250 mm·min^{-1}	0.5 mm 1500 mm·min^{-1}
Without boron	361.0 ± 27.5	537.0 ± 41.1	419.0 ± 74.1	523.0 ± 67.2	428.0 ± 15.3.0
With boron	596.0 ± 45.6	752.0 ± 37.4	1005.0 ± 104.1	979.0 ± 47.0	1772.0 ± 260.3

4. Conclusions

In the present research, the surface laser processing of 18Ni-300 maraging steel part manufactured by SLM process was investigated. The influence of the processing parameters and preposition of boron prior to the processing on the laser pool geometry and hardness of processed areas was studied. The following conclusions may be drawn based on the obtained results.

- The application of CO_2 continuous laser at the parameters of 1 kW laser power, 0.5 mm laser beam spot diameter and laser scanning speed in the range between 500 and 1250 mm·min^{-1} allows obtaining laser pool of acceptable geometry and sizes directly on as-manufactured

SLM part surface without any pre-processing. The increasing scanning speed to 1500 mm·min^{-1} or spot size to 2.0 and 3.0 mm results in too small pool depth and unstable pool geometry. Surface processing at 4.0 mm spot diameter provides only heating of the surface.

- The laser processing with preposition of amorphous boron layer provided surface alloying effect of the maraging steel SLM part – the hardness of processed surface areas ranged between ~600 HV0.2 at the lowest scanning speed and ~1770 HV0.2 at the highest speed, what is from 18% to ~3.5 times bigger, as compared with maximum hardness of 18Ni-300 maraging steel after aging (~58 HRC or ~510 HV). The geometry of laser pool did not differ significantly from that of laser-processed without boron.

- The laser boronising provided surface hardening effect; however, further research and parameters optimisation are required, which could allow solving the problem of appearance of transverse cracks and pores in the processed passes and understand better the effect of applied laser beam power density and heat input on the boron concentration in obtained pool and phase composition of the alloyed surface.

References

[1] Deckard, C. Method and apparatus for producing parts by selective sintering. U.S. Patent 4,863,538, filed October 17, 1986, published September 5, 1989.

[2] Maraging steels. In ASM Handbook, Volume 1: Properties and Selection: Irons, Steels, and High-Performance Alloys, 10th Edn. ASM International, Materials Park, Ohio, 1990. Vol. 1, pp 1869–1887

[3] Hallemans B., Wollants P., Roos I. R. Thermodynamic Reassessment and Calculation of the Fe-B Phase diagram. Zst. Metallkunde. 1994. Bd 85, N 10. S. 676-682.

[4] F. Cajner, D. Landek, S. Šolic, H. Cajner. Effects of thermochemical treatments on properties of maraging steels, Surface Engineering 2014 22 6, pp 468-471.

[5] Černašėjus, O.; Škamat, J.; Markovič, V.; Višniakov, N.; Indrišiūnas, S. Effect of Laser Processing on Surface Properties of Additively Manufactured 18-Percent Nickel Maraging Steel Parts. Coatings 2020, 10, 600.

[6] Behzad Fotovvati , Steven F. Wayne, Gladius Lewis, and Ebrahim Asadi. A Review on Melt-Pool Characteristics in Laser Welding of Metals. Advances in Materials Science and Engineering. 2018, ID 4920718.

[7] D. Bhaduria, P. Pencheva, A. Batala et al., "Laser polishing of 3D printed mesoscale components," Applied Surface Science, vol. 405, pp. 29–46, 2017.

[8] Balachninaitė, O., Bargelis, A., Dementjev, A., Jonušas, R., Račiukaitis, G., Sirutkaitis, V. 2008. Lazerinė technologija.Vilniaus universiteto leidykla. (in lithuanian)

[9] Chernov, Ya.B., Anfinogenov, A.I., and Shurov, N.I., Borirovanie stalei v ionnykh rasplavakh (Boronizing Steel in the Ionic Melts), Yekaterinburg: UrO RAN, 2001. (in russian)

IOP Conf. Series: Materials Science and Engineering 1239 (2022) 012010 doi:10.1088/1757-899X/1239/1/012010

Small size 2-DOF piezoelectric platform for unlimited locomotion

A Čeponis[1], D Mažeika[2], V Jūrėnas[3], P Vasiljev[4], R Bareikis[4], S Borodinas[5] and V Ostaševičius[3]

[1] Department of Engineering Graphics, Faculty of Fundamental Sciences, Vilnius Gediminas Technical University, Saulėtekio avn. 11, Vilnius, Lithuania

[2]Department of Information Systems, Faculty of Fundamental Sciences, Vilnius Gediminas Technical University, Saulėtekio avn., 11, Vilnius, Lithuania

[3]Institute of Mechatronics, Kaunas University of Technology, K. Donelaičio str. 73, Kaunas, Lithuania

[4]Education Academy, Vytautas Magnus University, K. Donelaičio str. 58, Kaunas, Lithuania

[5] Department of Applied Mechanics, Faculty of Civil Engineering, Vilnius Gediminas Technical University, Saulėtekio avn. 11, Vilnius, Lithuania

Abstract. The paper represents a numerical and experimental investigation of a 2-DOF piezoelectric positioning platform that can provide unlimited locomotion in the plane. The platform's operation principle is based on the excitation of the first bending mode of piezoelectric bimorph plates. A numerical investigation of the 2-DOF platform was performed, and it was found out that the suitable vibration mode of the bimorph plates occurs at 23.54 kHz Harmonic response analysis showed that the maximum displacement amplitude of the contact reached 561.12 µm or 2.67 $\mu m/V_{p-p}$ while an excitation signal of 210 V_{p-p} was applied. The experimental study was performed and showed that maximum linear velocity of 44.45 mm/s or 0.212 $mm/s/V_{p-p}$ were obtained while the load of 55.68g and applied and electrical signal of 210 V_{p-p} was used. In addition, at the same conditions output force of the platform was measured, and it was found that the platform is able to provide 44.16 mN of output force.

1. Introduction

For nova days, modern technologies used in biology, chemistry, physics of light, and other scientific high-resolution imaging systems have massive demand on positioning systems which are able to provide micro and nanoscale motion, positioning in several degrees of freedom, with fast response time and high dynamic characteristics [1-3]. There can be found numerous manipulation and positioning systems which operation are based on electromagnetic, electrostatic, and piezoelectric actuators [4-5]. Piezoelectric actuators have advanced features and able to provide high positioning resolution, have self-locking ability, fast response time, and are magnetic and static field free. Several degrees of freedom systems can be designed using a single piezoelectric transducer [3, 4]. Moreover, piezoelectric actuating devices do not require gearing mechanisms, therefore, no backlash appears during their operation. Usually piezoelectric positioning devices have a simple design based on regular structures so relatively simple driving and control methods, as well as operation electronics, is used [6].

Bansevičius *et al.* introduced a novel concept of multi-DOF actuators, which are able to provide planar and angular motions [7]. Proposed actuators are based on ring and cylinder-shaped transducers and have a simple design. It was shown that angular and planar resolutions of 20 µrad and 5µm can be achieved

Content from this work may be used under the terms of the Creative Commons Attribution 3.0 licence. Any further distribution of this work must maintain attribution to the author(s) and the title of the work, journal citation and DOI.
Published under licence by IOP Publishing Ltd

using cylinder type transducer while resolutions of 10μrad and 1μm resolutions, respectively was reached using ring type transducer.

Vasiljev *et al.* reported on a piezoelectric motor suitable for planar stage positioning XY plane [8]. The shaking beam principle was implemented to design a motor. It is composed of four transducers and a metal frame. Numerical and experimental studies revealed that planar motion resolution of the proposed motor is approximately 1.3μm in X and Y directions.

The aim of this research is to develop a novel design of an ultrasonic piezoelectric 2-DOF platform for unlimited high precision locomotion controlled by a single harmonic signal. Numerical and experimental studies were made, and the feasibility of the platform was validated. Results are discussed, and corresponding conclusions were made.

2. The design and operation principle of the platform

The platform design is based on three rectangular plates, six piezo ceramic plates, a triangle platform, and six thin junction beams. The rectangular plates act as a passive layer of the piezoelectric bimorphs which are formed by gluing piezo ceramic plates on both sides of the plates. The thin junction beams are used to compose piezo ceramic bimorphs and platform to one indissoluble structure. The junction beams are placed and the nodal lines of the first, out of the plane, bending mode of the bimorph plates (B_{20}) in order to reduce structural damping of the plates as well as to isolate vibrations of bimorphs from the platform. Moreover, at the bottom of each bimorph plate are placed sphere-shaped contacts which are used to transfer vibrations of the bimorph plates to the contact surface. The exploded and assembled view of the platform is shown in Fig.1.

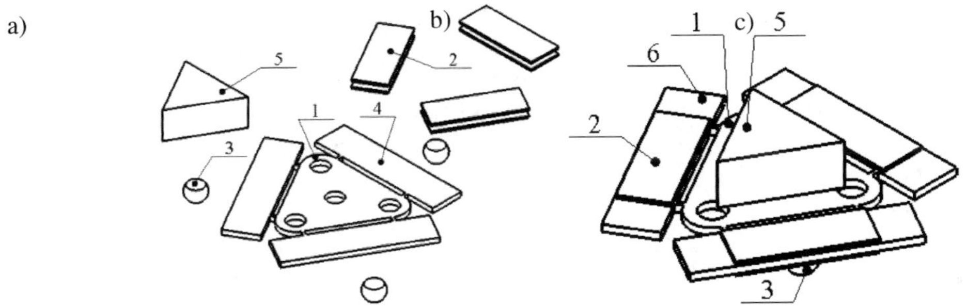

Figure 1. Design of the platform; a – exploited view; b – assembled view; c - prototype the platform; 1 – a platform; 2 – piezo ceramic plates; 3 – spherical contacts; 4 – rectangular plate; 5 – load; 6 – piezoelectric bimorph plate.

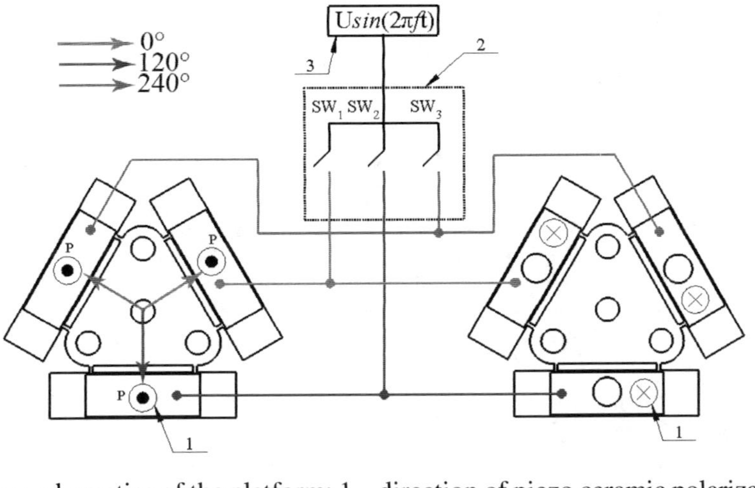

Figure 2 Excitation schematics of the platform; 1 – direction of piezo ceramic polarization; 2 – digitally controlled switch box; 3 – generator

IOP Conf. Series: Materials Science and Engineering 1239 (2022) 012010 doi:10.1088/1757-899X/1239/1/012010

The operation principle of the platform is based on the excitation of the first, out-of-plane bending mode (B_{20}), of the bimorph plate. Considering that bimorph plate clamping is unsymmetrical i.e. only one edge of the bimorph is fixed to the platform, then out-of-plane vibrations generate an additional displacement component in the Y direction due to unsymmetrical reaction forces. The excitation of the platform motion can be implemented by one harmonic signal, which must be applied to the appropriate piezoelectric bimorph plate. The excitation of one piezoelectric bimorph plate, while others are passive, will ensure the motion of the platform in a direction aligned to the transverse axis of the active bimorph. In addition, the motion direction of the platform can be controlled by switching excitation signals between different bimorph plates using a digitally controlled switch box (Fig.2). The trajectory planning algorithms can be applied in order to obtain the required planar motion of the platform.

3. Numerical investigation of the platform

The numerical investigations of the piezoelectric platform were performed in order to obtain optimal geometrical parameters, indicate electrical and mechanical characteristics as well as evaluate the proposed design. Modal analysis was made to identify natural frequencies and modal shapes of the platform. A frequency response study was performed to calculate impedance and phase-frequency characteristics as well as displacement–frequency characteristics. In addition, motion trajectories of the contacting spheres were analyzed to verify the operation principle. Finite element modeling of the piezoelectric platform was made using Comsol Multiphysics 5.4 software. The following materials were used in the model i.e. stainless steel for the passive layer of the platform and bimorph plates, PZT8 was used for piezo ceramic plates, and finally, alumina oxide ceramic was used for spherical contacts.
Results of the modal analysis revealed that the modal shape showed in Fig. 3a can be used to generate platform locomotion. It can be seen that the bimorph plates have B_{20} deformation mode while the platform body stays passive. Moreover, it can be found that B_{20} mode is modified by non-symmetrical clamping i.e. the bimorph plate not only bends in out of plane direction but also generates displacement in the Y direction. Therefore, the composition of these two displacement components will ensure the planar motion of the platform.

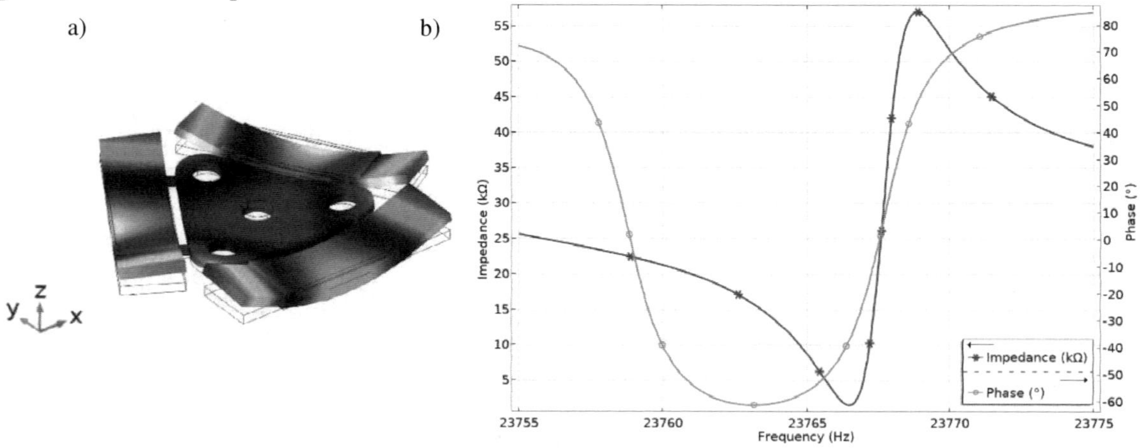

Figure 3. Electro-mechanical characteristics of the platform; a - the modal shape of the platform obtained at a frequency of 23.54 kHz; b - Impedance and phase-frequency characteristics in the frequency range of 23.755 – 23.775 kHz

The next step of the numerical investigation was dedicated to the investigation of the electrical characteristics of the platform. The impedance and phase-frequency characteristics of the platform are shown in Fig 3b.
The impedance graph confirms that the resonant frequency of the platform is obtained at 23.767 kHz (Fig. 3b). The minor mismatch of 220 Hz compared to the natural frequency occurs due to differences in electrical boundary conditions. The impedance value at the resonance frequency is 1.61 kΩ while the mechanical quality factor (Q_m) and effective coupling coefficient (k_{eff}) are 1398.1 and $12.9 \cdot 10^{-3}$, respectively.

The influence of the excited bimorph plate vibrations on the remaining bimorph was analyzed. It was important to evaluate parasitic vibrations of the non-excited plates. The bimorph plates were excited in accordance with the excitation schematics given in Fig. 2. The results are given in Fig. 4, and it can be seen that the contact point located on $Plate_1$ has a displacement amplitude of 561.12μm while contact points located on $Plate_2$ and $Plate_3$ plates have 56.45 μm and 48.19 μm displacement amplitudes, respectively. The difference between displacement amplitudes of the active plate and passive plates is about ten times.

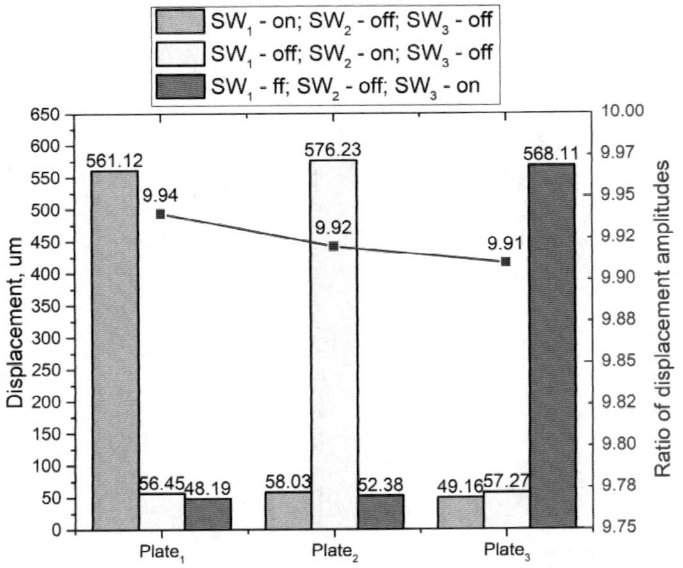

Figure 4. Summary of displacement amplitudes while different plates are excited.

It can be concluded that the excited vibrations of $Plate_1$ will have a minor effect on the vibrations of the passive plates and will not move the platform.

The motion trajectory of the contacting sphere was simulated when excitation voltage of $210V_{p-p}$ was applied to $Plate_1$. The time-domain study was set up for the same numerical model and boundary conditions. It can be seen that the trajectory is closed to the straight line. It indicates that the platform will be driven by implementing the vibro-impact principle. The projection length to the X-axis is 89.9μm, while the projection to Y-axis has a length of 551.1μm.

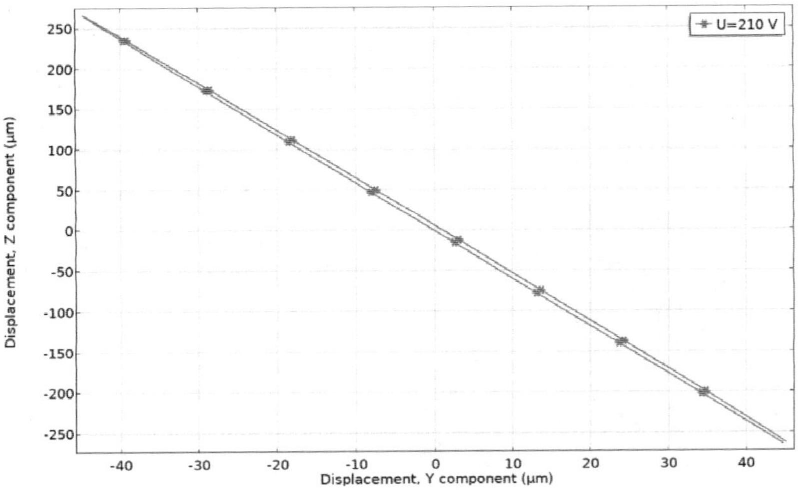

Figure 5. Motion trajectory of contact located on $Plate_1$, excitation voltage is $210V_{p-p}$

4. Experimental investigation of the platform

A prototype piezoelectric platform was made for the experimental study (Fig. 1c). The first stage of the experimental investigation was to measure impedance-frequency and phase-frequency characteristics of the platform. Measurement of each bimorph plate was done separately using impedance analyser.

Analyzing results, it can be noted that the are minor differences in resonant frequencies between bimorph plates (Fig.6). The resonant frequencies of $Plate_1$, $Plate_2$, and $Plate_3$ are 22.47 kHz, 23.04 kHz, and 22.86 kHz, respectively. The differences between resonant frequencies do not exceed 570Hz. Moreover, compared to results obtained during numerical investigation difference between resonant frequencies is 1.29 kHz, respectively. The linear speed of the platform was measured when the different excitation voltage and the preload were applied. Values of maximum linear speed at different load conditions are given in Fig. 7a. The characteristics were measured by digital tachometer DT-2100 (USA) with linear speed measuring function. Fig. 7a shows that the highest platform motion velocities were obtained at $210V_{p\text{-}p}$ excitation voltage for all load and plate excitation cases. The results found that the highest linear motion values were obtained while the load was equal to 55.68g, and it reached 44.45 mm/s or 0.182 $mm/s/V_{p\text{-}p}$. Moreover, it was found that velocity values at different loads are similar and will ensure even linear motion speed in all directions.

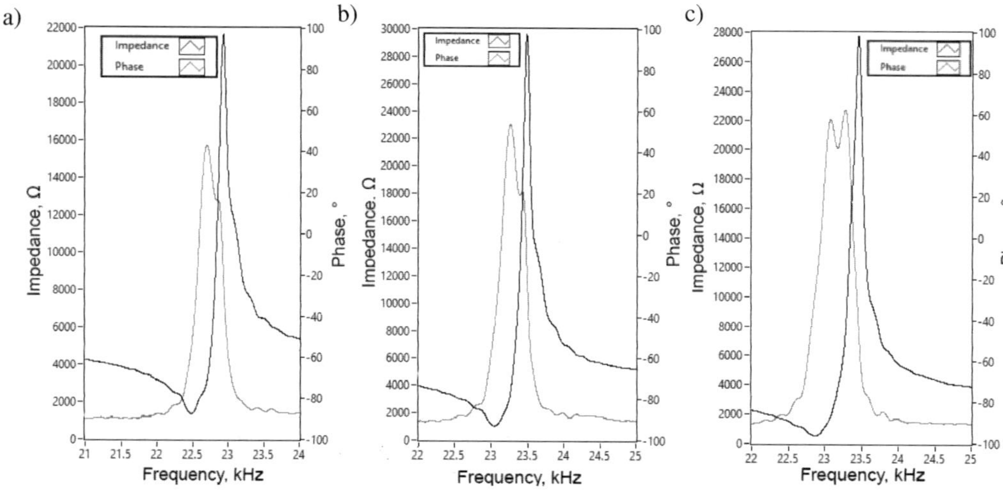

Figure 6. Impedance and phase frequency characteristics of the platform; a – characteristics of $Plate_1$; b – characteristics of $Plate_2$; c – characteristics of $Plate_3$

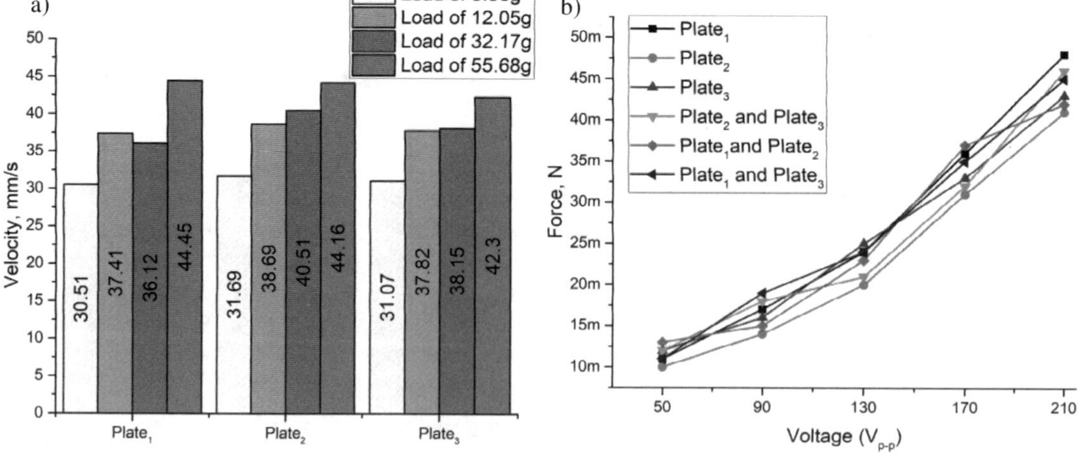

Figure 7. Characteristics of the platform; a – maximum velocity values; b – output force characteristics

The next stage of the experimental investigation was dedicated to an indication of output force characteristics of the platform when the load was set to 32.17g, and different excitation voltages were applied in the range from $50V_{p-p}$ to $210V_{p-p}$. The measurements were performed by an analogical force gauge ATN 0.05 (China). Analysing results, it can be found that the highest output force was obtained at $210Vp_{-p}$ while the average output force is 44.16 mN (Fig.7b). The deviation between values is 34.83μN. Finally, the ratio of the average output force to excitation voltage is $210\mu N/V_{p-p}$. Therefore, on the basis of these results can be found that the output force increment relationship to an increment of excitation voltage is almost linear and well predictable.

5. Conclusions

2-DOF planar motion piezoelectric platform was developed and investigated. The design of the platform is simple and well scalable. The design of the platform is based on three piezoelectric bimorph plates, which were evenly distributed around a triangle-shaped platform. The platform's design provides partial isolation of parasitic vibrations transfer from active plate to passive ones as well as to the platform body. Numerically calculated impedance and phase frequency characteristics showed that the operation frequency of the platform is 23.76 kHz. These results also confirmed results obtained during modal analysis of the platform. Numerical investigations of the platform showed that parasitic vibrations of the unexcited plates are low, and their influence on the planar motion is negligible. Also, it was found that the maximum displacement amplitude of the contact reached 561.12 μm or 2.67 $\mu m/V_{p-p}$ .Experimental investigation of dynamic characteristics showed that that average planar motion reached 44.45 mm/s or 0.212 $mm/s/V_{p-p}$ while excitation voltage amplitude was $210V_{p-p}$ and preload was 55.68g while maximum output force reached 44.16 mN at the same conditions.

Acknowledgment

This research has received funding from the European Regional Development Fund (project No 01.2.2-LMT-K-718-01-0010) under a grant agreement with the Research Council of Lithuania.

6. References:

[1] Wu, Z., Xu, Q., 2018, Survey on recent designs of compliant micro-/nano-positioning stages, Actuators 4 (1) doi:10.3390/act7010005

[2] Mohith, S., Upadhya, A.R., Karanth, N.P., Kulkarni, S.M., Muralidhara, M.Rao., 2021 Recent trends in piezoelectric actuators for precision motion and their applications: a review, Smart Materials and Structures, 30 (1), doi: 10.1088/1361-665X/abc6b9

[3] Bansevicius, R. 2009. State-of-the-art and new developments of multi-degree-offreedom piezoelectric motors for experimental mechanics and measuring devices. 19th IMEKO World Congress 2009, 1(January 2009) pp. 315-319. ISBN 9781615675937.

[4] Liu, Y., Deng, J., Su, Q. 2018. Review on Multi-Degree-of-Freedom Piezoelectric Motion Stage, 8, pp. 59986-60004, doi: 10.1109/ACCESS.2018.2875940

[5] Ding, N., Zou, J., Gu, G., 2020. Design of dielectric-elastomer actuated XY stages with millimeter range and submicrometer resolution, Proceedings of MARSS 2020: International Conference on Manipulation, Automation, and Robotics at Small Scales, Issue: 2019, pp. 2-7, 10.1109/MARSS49294.2020.9307920

[6] Gao, X., Yang, J., Wu, J., Xin, X., Li, Z., Yuan, X., Shen, X., Dong, S. 2020. Piezoelectric Actuators and Motors: Materials, Designs, and Applications, Advanced Materials Technologies, 5(1) pp. 1-26. doi: 10.1002/admt.201900716

[7] Bansevičius, R., Mažeika, D., Jūrėnas, V., Kulvietis, G., Drūkteinienė, A. 2019. Multi-DOF Ultrasonic Actuators for Laser Beam Positioning. Shock and Vibration. doi: 10.1155/2019/4919505

[8] Vasiljev, P., Borodinas, S., Yoon, S. J., Mažeika, D., Kulvietis, G. 2005. The actuator for micro moving of a body in a plane. Materials Chemistry and Physics. 91(1) pp. 237-242. doi: 10.1016/j.matchemphys.2004.09.040

Mathematical modelling and theoretical research of micropositioning system

Jurga Subačiūtė - Žemaitienė*, Vaiva Trečiokaitė, Nikolaj Šešok, Ernestas Šutinys, Andrius Dzedzickis, Justas Tamošiūnas, Viktor Masalskyi and Vytautas Bučinskas

Vilnius Gediminas technical university, Department of Mechatronics, Robotics and Digital manufacturing, J. Basanaviciaus str. 28, LT-03224 Vilnius, Lithuania

*Corresponding author e-mail: jurga.subaciute-zemaitiene@vilniustech.lt

Abstract. In this paper, modelling and theoretical research of micropositioning system with stepper motor are proposed. The aim of our research is to provide a method, which is suitable for modelling a ball-screw based positioning system controlled by a stepper motor. A created dynamical model of micropositioning system mechanical parts. The second type of Lagrange equation is used to construct the mathematical model. For modelling micropositioning systems, Matlab/SIMULINK software is used. MatLab is used to perform calculations of some equations, and SIMULINK is used to create a dynamic model of a micropositioning system and simulate system operation. Analysis of motion smoothness showed that even for systems with relatively small screw radius and pitch 1/4 microstepping mode are unable to provide smooth movement. The smoother the movement, the less energy is required to overcome friction.

Keywords: Mathematical modelling micropositioning system

1. Introduction

Robotic micropositioning systems - devices that operate with micrometric accuracy and are used in industry and scientific research [1–3]. Micropositioning systems are able to increase the adaptability of the systems using them by performing small, highly precise movements. Due to the rapid development of technologies and the wide range of applications, the demand for micropositioning systems is growing every year [1, 4]. This is especially relevant for microbiological studies, which require the accuracy of micrometres. The choice of which positioning system will be needed to perform the desired tests is based on the information provided in the specifications of the analyzed equipment: workflow, resolution, accuracy, error rate. The majority of micropositioning systems suffer from transient processes that affect the motion's resolution, accuracy, and smoothness [5, 4].

With the increasing demand for micropositioning systems, ball-screw based systems driven by stepper motors or servo drives are among the most popular choices for high positioning accuracy and repeatability. One of the methods to reduce the impact of transient processes is the mathematical modelling of such systems [6, 7]. Simulation of systems behaviour under various operational conditions saves time required for their development and allows implementing modern control methods. To achieve the highest possible positioning accuracy of the system, machine learning can be used [8–11]. Many artificial intelligence developers agree that it is easier to train a system by showing its input and output

Content from this work may be used under the terms of the Creative Commons Attribution 3.0 licence. Any further distribution of this work must maintain attribution to the author(s) and the title of the work, journal citation and DOI.
Published under licence by IOP Publishing Ltd

behaviour by predicting the desired response to all implied inputs than programming them manually [12] since mathematically obtained results allow to determine system weaknesses and chose suitable control and positioning errors compensation methods [13, 14].

The research aims to provide a method suitable for modelling a ball-screw based positioning system controlled by a stepper motor. The created model considers all main issues: characteristics of a mechanical system and stepper motor control features.

2. Modelling of micro positioning system

In our case, we are modelling the x-axis of a 3-axis micropositioning system intended for measurements of biological samples. The kinematic diagram of the modelled system provided in Figure 1. The mechanical structure of the system x-axis consists of 1 - stepper motor, 2 - coupling, 3 - micrometric accuracy ball screw, 4 - nut and 5 - working table. Ball screw drive converts the rotary motion of the stepper motor shaft into linear movement of the working table.

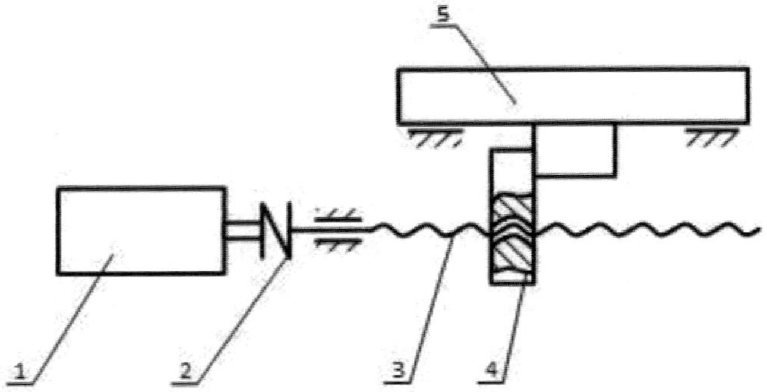

Figure 1. Kinematic diagram of system x-axis.

The modelling of a mechanical part of a system begins with the creation of its dynamic model Figure 2. Torque M, generated by the stepper motor, creates the motion of the mechanical system. The rotor of the stepper motor, ball screw, nut and working table is considered as rigid bodies and their deformation due to acting load, and thermal expansions are neglected.

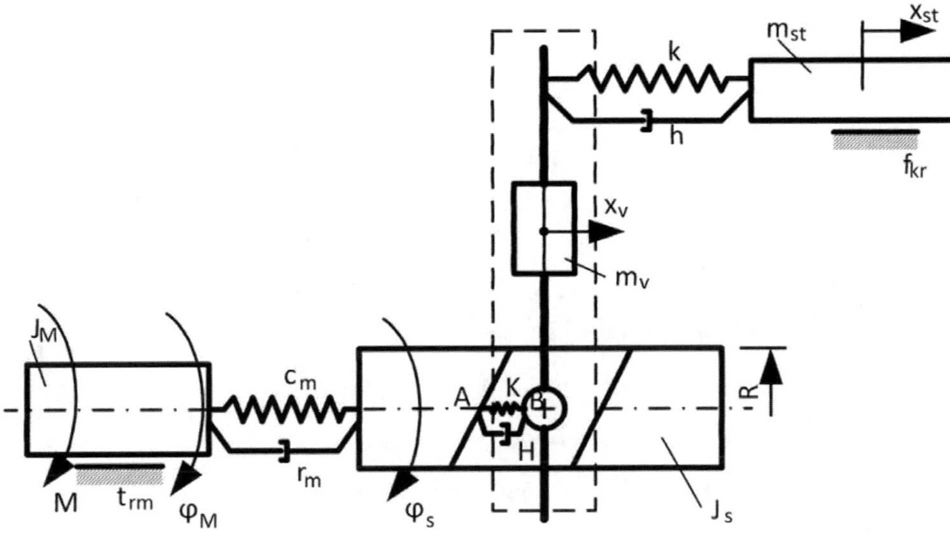

Figure 2. Dynamic model.

IOP Conf. Series: Materials Science and Engineering 1239 (2022) 012011 doi:10.1088/1757-899X/1239/1/012011

In the model evaluated moment of motor rotor inertia J_M, moment of screw inertia J_S, nut mass m_v, and mass of the working table m_{st}. The coupling, connecting the motor shaft to the ball screw, is modelled as an elastic element with damping. Rotational stiffness and damping coefficients are noted as c_m and r_m.

The contact between the nut and the table here modelled as an elastic element with damping, which stiffness and damping coefficients are *k* and *h*. It is assumed that the mechanical contact between the rolling balls, balls screw, and the nut can be modelled as a single elastic element with damping, with a total stiffness coefficient *K* and damping coefficient *H*. This element is shown in the dynamic model (Figure 2) between points *A* and *B*. Point *A* belongs to the ball screw and point *B* to the nut. Such an approach show that the ball and nut form a single body in the model.

Friction in the linear guideways of the working table – damping t_{kr}, and in stepper motor bearings – damping t_{rm}, is also evaluated, as well as, the effect of electromagnetic force generating rotor motion in the stepper motor.

The system's current state described using the generalized coordinates: φ_M – rotation angle of the motor rotor, φ_s – rotation angle of the ball screw, x_v – displacement of the nut, x_{st} – displacement of the working table.

Mathematical model of the system derived using second-order differential equations of the Lagrange method:

$$\frac{d}{dt}\left(\frac{\partial T}{\partial \dot{q}_i}\right) - \frac{\partial T}{\partial q_i} + \frac{\partial \Phi}{\partial \dot{q}_i} + \frac{\partial \Pi}{\partial q_i} = F_i(t) \tag{1}$$

where T, Π – the kinetic and potential energy of the system; Φ – dissipative function; F_i(t) – external generalized force acting in the direction described by coordinate q_i.

The kinetic energy, the potential energy, and the dissipation function of the dynamic model can be expressed using the following equations:

Kinetic energy:

$$T = \frac{1}{2}J_M \cdot \dot{\varphi}_M^2 + \frac{1}{2}J_S \cdot \dot{\varphi}_S^2 + \frac{1}{2}m_v \cdot \dot{x}_V^2 + \frac{1}{2}m_{st} \cdot \dot{x}_{st}^2 \tag{2}$$

Potential energy:

$$\Pi = \frac{1}{2}c_m(\varphi_M - \varphi_s)^2 + \frac{1}{2}K\Delta^2 + \frac{1}{2}k(x_v - x_{st})^2 \tag{3}$$

Displacement of point A:

$$\mathrm{R} \cdot \varphi_s \cdot \mathrm{tg}\alpha \tag{4}$$

Where R – screw working radius; α – helix angle.

Displacement of point B is equal to the displacement of nut x_v.

Thus:

$$\Delta = (R \cdot tg\alpha) \cdot \varphi_s - x_v \tag{5}$$

Dissipative function:

$$\Phi = \frac{1}{2}r_m(\dot{\varphi}_M - \dot{\varphi}_s)^2 + \frac{1}{2}t_{rm} \cdot \dot{\varphi}_M^2 + \frac{1}{2}H\dot{\Delta}^2 + \frac{1}{2}h(\dot{x}_v - \dot{x}_{st})^2 + \frac{1}{2}t_{kr} \cdot \dot{x}_{st}^2 \tag{6}$$

Here:

$$\dot{\Delta} = (R \cdot tg\alpha) \cdot \dot{\varphi}_s - \dot{x}_v \tag{7}$$

Then derived model of the mechanical system expressed as a system of equations:

$$\begin{cases} J_M \cdot \ddot{\varphi}_M + (r_m + t_{rm})\dot{\varphi}_M - r_m\dot{\varphi}_s + c_m \cdot \varphi_M - c_m \cdot \varphi_s = M \\ J_S \cdot \ddot{\varphi}_s - r_m \cdot \dot{\varphi}_M + (H \cdot R^2(tg\alpha)^2 + r_m)\dot{\varphi}_s - HR(tg\alpha)x_v - c_m \cdot \varphi_M + \\ +(K \cdot R^2(tg\alpha)^2 + c_m)\varphi_s - KR(tg\alpha)x_v = 0 \\ m_v\ddot{x}_v - HR(tg\alpha)\dot{\varphi}_s + (H+h)\dot{x}_v - h \cdot \dot{x}_{st} - KR(tg\alpha)\varphi_s + (K+k)x_v - \\ -k \cdot x_{st} = 0 \\ m_{st}\ddot{x}_{st} - h\dot{x}_v + (h + f_{kr})\dot{x}_{st} - kx_v + kx_{st} = 0 \end{cases} \quad (8)$$

Derived system of equations is solved in respect of the second derivative of generalized coordinate $\ddot{q}_i$.

The mechanical system receives the excitation from the torque generated by the electromagnetic force on the stepper motor output shaft:

$$M = \sum_{j=1}^{m} M_j \quad (9)$$

Here M – torque on the rotor generated by the motor winding; M_j – torque on the stator generated by winding; m – number of windings.

Torque on the stator generated by a single winding

$$M_j = k_m \cdot \sin(n\varphi_M(t) + \varphi_0) \cdot I_j(t) \quad (10)$$

Here n – number of stepper motor poles; φ_M – rotation angle of stepper motor rotor (rad); I_j – current in the winding (A).

Current in the winding derived from:

$$U_j = emf_j + R_v \cdot I_j(t) + L\frac{dI_j(t)}{dt} \quad (11)$$

Here U_j – winding voltage (V); emf_j – reversible electric drive (V); R_v – winding resistance (Ω); L – winding inductance.

Reversible induced electromagnetic force evaluated as follows:

$$emf_j = k_m \cdot \sin(n\varphi_M + \varphi_o) \cdot \dot{\varphi}_M \quad (12)$$

k_m – motor constant (A/Nm); $\dot{\varphi}_M = \omega_M$ – rotor angular speed (rad/s).

For modelling, micropositioning system Matlab/SIMULINK software is used. The simplified block diagram provided in 'Figure 3'. To simulate the operation of the stepper motor, a control signal generator block is used to control the torque. It simulates supply voltage U_1 and U_2 applied to the winding of the stepper motor. For this purpose, we used a harmonic signal which, is digitized according to microsteping mode, which we are modelling.

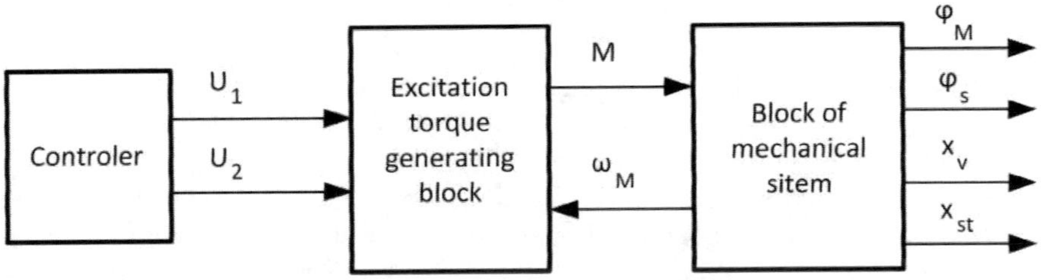

Figure 3. Simplified block diagram of micropositioning system.

3. Methodology of the theoretical research

One of the main goals of modelling was to evaluate the influence of the stepper motor control mode on the overall system characteristics and to determine the minimum required micro-stepping mode. To fulfil the aim, simulated behaviour of the system when working table moves with the averaged speed of 35 μm/s and the stepper motor is controlled in various modes. As input signals for the model were generated 4 different datasets representing input voltages (Figure 4) when the motor rotates 11.5 degrees

IOP Conf. Series: Materials Science and Engineering 1239 (2022) 012011 doi:10.1088/1757-899X/1239/1/012011

and is operated in 1/4, 1/16, 1/64 and 1/256 microstepping modes. Other model parameters provided in table 1.

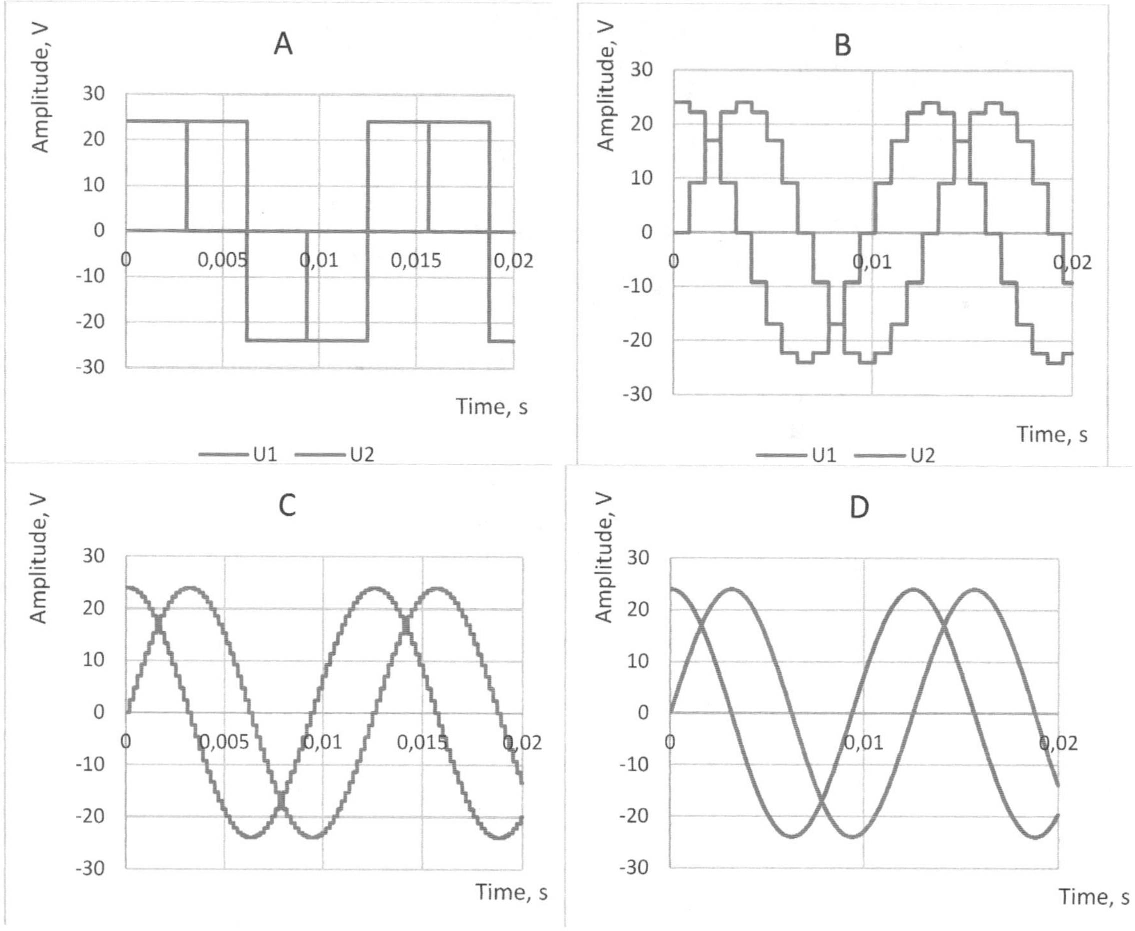

Figure 4. Four types of input signals. Each graph corresponds to a different stepper motor operation mode: A – microstepping is 1/4, B – microstepping is 1/16, C – microstepping is 1/64, D – microstepping is 1/256.

The smoothness of motion of the stepping motor output shaft depends on the type of used microstepping mode and the load acting on the output shaft. The inertia of the mechanical parts could smoothen the motion if the system is not very stiff. Nevertheless, the lack of stiffness in system components can have a negative impact on its overall accuracy and performance. From obtained input signals, it is obvious that 1/4 stepping mode will generate the most uneven motion, 1/256 stepping mode provides smooth control voltage similar to pure sinewave, but unfortunately, microstepping reduces the motor torque. Therefore, to ensure optimal system performance and to have the possibility to implement modern positioning accuracy increase methods[15, 16], it necessary to select the most appropriate motor operating mode.

Table 1. Model input data.

Parameter	Value
Stiffness coefficient of contact between nut and table – k, N/m	1e9
Damping coefficient of contact between nut and table – h, N*s/m	70
Table weight – m_{st}, kg	1.5
Nut weight – m_v, kg	0.05
Screw-nut drive stiffness coefficient – K, N / m	4.7e6
Damping koefficient of the screw-nut drive – H, N * s/m	5000
Coupling stiffness coefficient – c_m, N*m/rad	1.5e3
Coupling damping coefficient – r_m, N*m*s/rad	15
Damping coefficient evaluating the friction of the stepper motor bearings and the action of electromagnetic forces – t_{rm}, N*s/m	0.0075
Screw moment of inertia – J_s, kg*m^2	3.56e-07
Screw working radius – R, m	4e-3
Screw line pitch angle – t_{galfa}	0.0995
Stepper motor moment of inertia – J_m, kg*m^2	0.036e-4
Damping coefficient of table guides – f_{kr}, N*s/m	10
Winding inductance – L_{ind} , mH	0.0023
Motor constant – k_{m1}, A/Nm	0.067
Motor constant – k_{m2}, A/Nm	0.067
Winding resistance – R, Ω	2.4
Number of poles of a stepper motor –N_{ra}, N_{rb}	50

4. Results

The most important results obtained during the modelling procedure presented in Figures 5 and 6. Figure 5 represents the rotational displacement of the screw φ_s (Figure 2), Figure 6 illustrates the movement of the working table x_{st} (Figure 2) when four different modes for stepper motor control are applied.

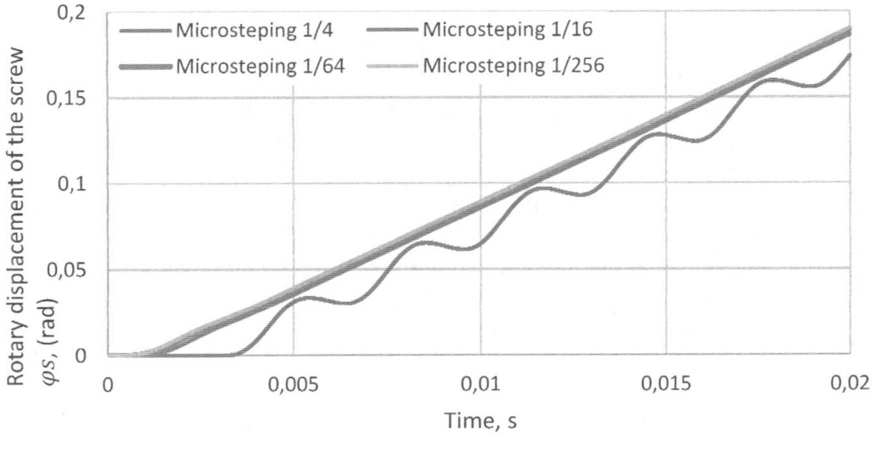

Figure 5. Angular displacement of the screw φ_s

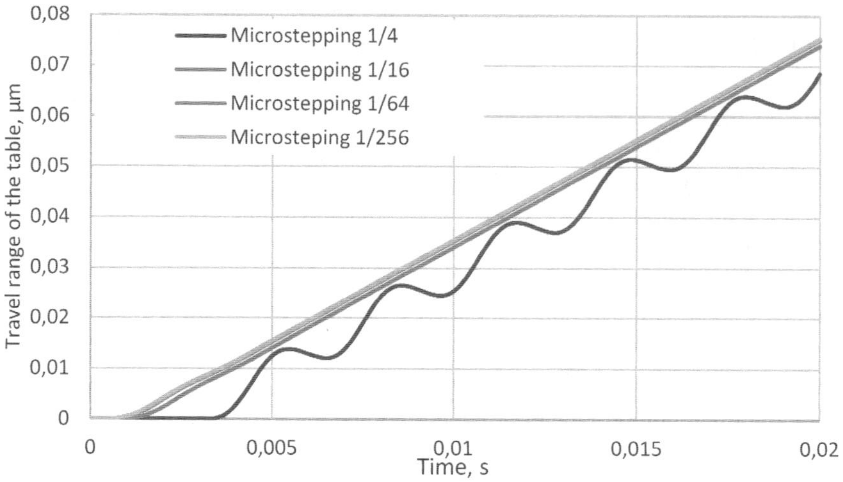

Figure 6. Displacement of the working table x_{st}

Analyzing screw and working table movement, it was found that they perform identical motions, just amplitudes are different. Such a result allows us to state that our mechanical system has an optimal design. There is no backlash, system stiffness and damping ratio is sufficient, inertia force and external excitation from the motor do not cause vibrations in the system.

Assessing the smoothness of the carriage motion, which is essential for systems operating with biological objects, found that the required velocity of 35 μm/s stepping mode 1/4 is insufficient. It is clearly noticeable small overshoot and short stops after each micro step of the rotor. Analyzing other microstepping modes, no significant differences were observed except a small offset in the travelled distance, which in the real system corresponds to its positioning accuracy. The work of internal frictional forces can explain the observed effect. In the case of unsmooth movement, more energy is wasted to overcome the friction of the system then it stops for a moment after each microstep.

5. Conclusions

Performed research provided perspective findings useful for the mathematical modelling of the ball screw based drives controlled by stepper motors.

In our research, we developed a mathematical model of a linear positioning system based on a micrometric precision ball-screw drive controlled by a stepper motor. Our model takes into account the main characteristics of mechanical systems components such as mass, stiffness, damping and friction between movable components. It also evaluates electromagnetic force acting in stepper motor and suits to simulate its behaviour under various applied voltages.

Obtained modelling results prove the correctness of the mechanical design of our modelled positioning system. No backlashes, oscillations or transitional processes revealed themselves during simulating of system operation at the velocity of 35 μm/s.

Analysis of motion smoothness showed that even for systems with relatively small screw radius and pitch 1/4 microstepping mode are unable to provide smooth movement. According to our analysis outcome, the optimal microstepping mode is 1/16 or 1/64. Nevertheless, taking into account accuracy requirements, 1/256 microstepping mode looks most promising, even motor rotation speed became limited by the power supply frequency. The smoother the movement ensure smaller energy losses due to overcoming static friction force.

The obtained finding also forms guidelines for our future research. It is necessary further validation of developed model by comparing its results with experimental results. After that, we are planning to implement machine-learning algorithms in our model to open wider possibilities of optimal parameter definition and control features optimization.

Acknowledgements
This project has received financial support from the Research Council of Lithuania (LMTLT), Nr. P-LLT-21-6, State Education Development Agency of Latvia, Ministry of Science and Technology (MOST) of Taiwan, Nr. MOST 110-2923-E-011-006-MY3.

References

[1] J. Juuti, M. Leinonen, and H. Jantunen, "Micropositioning," pp. 319–320, 2008.

[2] H. XI *et al.*, "+ LJK SUHFLVLRQ : DIHU OHYHO & KLS' HWHFWLRQ," pp. 381–387, 2016.

[3] H. XI *et al.*, "A novel flexure-based XY motion compensator: Towards high-precision wafer-level chip detection," *2016 IEEE 18th Electron. Packag. Technol. Conf.*, pp. 381–387, 2016.

[4] S. Tayouri, I. Izadi, and J. Ghaisari, "Modeling and Parameter Identification of Piezoelectric Actuator in Micropositioning Systems," *ICEE 2019 - 27th Iran. Conf. Electr. Eng.*, pp. 1193–1198, 2019, doi: 10.1109/IranianCEE.2019.8786431.

[5] K. Poletkin, "Static Pull-in Behavior of Hybrid Levitation Micro-Actuators: Simulation, Modelling and Experimental Study," *IEEE/ASME Trans. Mechatronics*, vol. 4435, no. c, pp. 1–11, 2020, doi: 10.1109/TMECH.2020.2999516.

[6] X. H. Nguyen, T. H. Mau, I. Meyer, B. L. Dang, and H. P. Pham, "Improvements of piezo-actuated stick-slip micro-drives: Modeling and driving waveform," *Coatings*, vol. 8, no. 2, pp. 1–16, 2018, doi: 10.3390/coatings8020062.

[7] F. Najar, S. Choura, S. El-Borgi, E. M. Abdel-Rahman, and A. H. Nayfeh, "Modeling and design of variable-geometry electrostatic microactuators," *J. Micromechanics Microengineering*, vol. 15, no. 3, pp. 419–429, 2005, doi: 10.1088/0960-1317/15/3/001.

[8] E. Bejar and A. Moran, "Deep reinforcement learning based neuro-control for a two-dimensional magnetic positioning system," *Proc. - 2018 4th Int. Conf. Control. Autom. Robot. ICCAR 2018*, pp. 268–273, 2018, doi: 10.1109/ICCAR.2018.8384682.

[9] M. Zhou, Y. Yu, J. Zhang, and W. Gao, "Iterative Learning and Fractional Order PID Hybrid Control for a Piezoelectric Micro-Positioning Platform," *IEEE Access*, vol. 8, pp. 144654–144664, 2020, doi: 10.1109/ACCESS.2020.3014725.

[10] J. Xu, K. Qin, Y. Xu, and W. Ji, "Method Combining Machine Vision and Machine Learning for Reed Positioning in Automatic Aerophone Manufacturing," *2019 4th Int. Conf. Robot. Autom. Eng. ICRAE 2019*, pp. 140–147, 2019, doi: 10.1109/ICRAE48301.2019.9043784.

[11] M. Leroux, M. Raison, T. Adadja, and S. Achiche, "Combination of eyetracking and computer vision for robotics control," *IEEE Conf. Technol. Pract. Robot Appl. TePRA*, vol. 2015-Augus, 2015, doi: 10.1109/TePRA.2015.7219692.

[12] M. I. Jordan and T. M. Mitchell, "Machine learning: Trends, perspectives, and prospects," vol. 349, no. 6245, 2015.

[13] H. N. Rahimi and M. Nazemizadeh, "Dynamic analysis and intelligent control techniques for flexible manipulators: A review," *Adv. Robot.*, vol. 28, no. 2, pp. 63–76, 2014, doi: 10.1080/01691864.2013.839079.

[14] D. V. Sabarianand, P. Karthikeyan, and T. Muthuramalingam, "A review on control strategies for compensation of hysteresis and creep on piezoelectric actuators based micro systems," *Mechanical Systems and Signal Processing*, vol. 140. Academic Press, p. 106634, Jun. 2020, doi: 10.1016/j.ymssp.2020.106634.

[15] M. Sumanas *et al.*, "Implementation of Machine Learning Method for Positioning Accuracy Improvement in Industrial Robot," *2020 IEEE Open Conf. Electr. Electron. Inf. Sci. eStream 2020 - Proc.*, no. 09, 2020, doi: 10.1109/eStream50540.2020.9108858.

[16] R. Szewczyk, *Advances in Intelligent Systems and Computing 1140 Automation 2020 : Towards Industry of the Future*. 2020.

IOP Conf. Series: Materials Science and Engineering 1239 (2022) 012012 doi:10.1088/1757-899X/1239/1/012012

An optimisation problem for exoskeleton-aided functional rehabilitation of an upper extremity

P Falkowski[1,2]
[1] Warsaw University of Technology, Warsaw, Poland
[2] ŁUKASIEWICZ Research Network – Industrial Research Institute for Automation and Measurements PIAP, Warsaw, Poland

Abstract. With pandemics, a problem of non-guaranteed constant access to rehabilitation aroused. As civilisation diseases attack a significant share of the population, providing motor treatment seems to be one of the biggest challenges towards the healthcare system of the future. While taking into account the severe level of disabilities and unexpected circumstances, the trainings may require being held in a house of a patient. Moreover, due to emergencies, a physiotherapist may not be able to participate in it personally. Thus, to mitigate the repercussions of such situations, advanced mechatronic systems for remote-home-rehabilitation have to be developed. Moreover, surveyed physiotherapists, claim that function-oriented training, robot aid, remote treatment and home therapy are the leading trends for rehabilitation of the future. Combining these is relatively difficult, as automated robotic rehabilitation requires an advanced control system. As to partially support motion along the pre-programmed trails, an optimisation problem needs to be solved. Not only to follow the trail transitionally but also to control a tilt of a gripped object and to avoid immediate large changes of velocities and accelerations if not necessary. This paper presents an approach towards an optimisation problem based on an exoskeleton of three driven and two free degrees of freedom. Simulated function-oriented training consists of diverse typical motions, such as drinking, eating, writing, steering a control wheel, lifting a bag from the ground, and lifting an object onto a shelf. Based on these, an optimisation problem is formulated. Works presented in the paper are an integral part of a planned long-term research on a home-remote-rehabilitation with a use of an exoskeleton.

1. Introduction

The COVID-19 pandemics exposed unnoticed weaknesses of different fields – urban planning, learning, social interactions, industry, sports, and medicine, among others. Some of these were intensively developed within an ongoing crisis; however, not all with satisfying outcomes [1].

As for now, rehabilitation is mostly provided by therapists who need to stay in the same room with patients. Therefore, it is not resistant to limitations related to crises. During the lockdown, many people under motor treatment were left without access to therapy. Thus, providing an access to constant rehabilitation seems to be one of the biggest challenges towards the healthcare system of the future.

Such a developed remote solution will not only be usable in emergency times. On a daily basis, patients living far from specialist clinics or the ones with a severe level of disability may use its advantage to work out in their houses. Moreover, the rehabilitation process may be organised with the best physiotherapists or medics from any part of the world [2].

It is highly possible that the COVID-19 pandemics will be followed by another one. Also, many post-COVID patients suffer from neurological diseases and limited performance [3]. Moreover, a correlation between being infected with SARS-CoV-2 and getting a stroke has been deduced [4]. Therefore, an investment in the technology of remote rehabilitation is not only prevention from the next crisis [5]. It

Content from this work may be used under the terms of the Creative Commons Attribution 3.0 licence. Any further distribution of this work must maintain attribution to the author(s) and the title of the work, journal citation and DOI.
Published under licence by IOP Publishing Ltd

IOP Conf. Series: Materials Science and Engineering 1239 (2022) 012012 doi:10.1088/1757-899X/1239/1/012012

is also fighting the repercussions of the ongoing one and strengthening the medical sector via multiplying the capabilities of therapists with the use of mechatronic devices.

1.1. Major challenges towards rehabilitation of the future

As mention before, rehabilitation of the future should be resistant towards changes in an environment above all. Moreover, questionnaire research was conducted with a group of 130 Polish physiotherapists to define requirements towards a designed exoskeleton-aided therapy. Nearly 80% of them mention a changed style of rehabilitation to the task-oriented as a leading trend for the future. It is particularly important, as 86% of respondents claim to use multi-joint mobilisation for kinesiotherapy. Moreover, 82% of physiotherapists would like to have an access to new knowledge on physiotherapy, and 60% of them would benefit from having an effective method of measuring the effectiveness of treatment.
A pandemic-proof solution should be dedicated to home use without a presence of a therapist onsite. 55% of questioned professionals chose remote or home training as an irreplaceable part of rehabilitation in 10 years. Thus, their answers were assessed in terms of compliance to devices enabling such a treatment. Over 62% of therapists' demands would be met by a home-use rehabilitation robot, while over 36% may be as well after enhancing it with additional technologies. As a result of requirements and limitations analyses, a rehabilitation exoskeleton with additional technology modules was selected as the most promising for this purpose.

1.2. Concept of a function-oriented rehabilitation of an extremity

A task-oriented rehabilitation is an alternative to the commonly practised manual therapy and single-joint mobilisation. Compared to these, it is superior as it activates multiple degrees of freedom (DOFs) at the same time; which particularly effective in post-stroke treatment [6]. Moreover, its aim is not only to restore mobility in joints and strengthen motor units but also to relearn genuine motion patterns and correlate them with appropriate neurological triggers. Therefore, such training consists of repetitions of real-life movements, particularly important for a patient; e.g. a professional driver may focus mainly on movements necessary to drive a car. Hence, a rehabilitated person is usually more motivated to work out and achieves results faster. As the treatment engages many joints from the beginning, it may require force support to weaker motor groups.

2. Construction of an exoskeleton for task-oriented motor treatment

An exoskeleton consists of five DOFs, among which three are driven and two are left free. Their denotation is presented in Table 1. A base position of a device when all DOFs' rotations are equal 0° is reached for a position of an extremity presented in Figure 1***Figure 3***. All trajectories are described in reference to a centre of a shoulder joint (centre of a coordinate system) as a motion of a characteristic point, a centre of a hand.

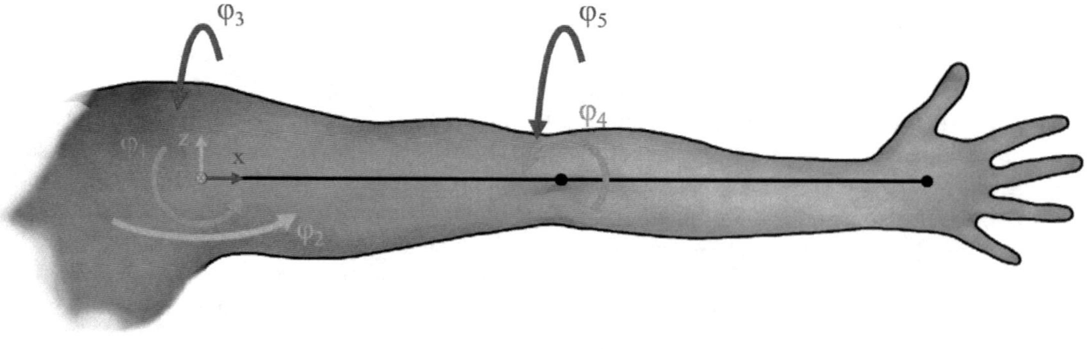

Figure 1. Base position of an extremity, corresponding to the base position of an exoskeleton; with rotations and a coordinate system for trajectory description

Table 1. DOFs of a modelled exoskeleton

Rotation	Joint	Motion	Driven/Free	min./max.
φ_1		Abduction (+) / adduction (-)	Driven	-140° / 90°
φ_2	Shoulder	Horizontal adduction (+) / abduction (-)	Driven	-50° / 130°
φ_3		Flexion (+) / extension (-)	Free	-50° / 130°
φ_4	Elbow	Flexion (+) / extension (-)	Driven	0° / 140°
φ_5		Supination (+) / pronation (-)	Free	-80° / 80°

Considered control over the exoskeleton combines user intentions detected by a sensory system, pre-calculated inverse dynamics for a desired trajectory and physical parameters of a device with an extremity of an average man (M50 percentile), and the main control algorithm. The last one is based on a predictive system with an AI-linearized model of an exoskeleton, and a set of rules activated only while an incorrect motion pattern is detected. Such an approach should limit possible motor compensation of movements and increase the safety of a patient.

As a physiotherapist cannot fully observe the patient, the device has to track their performance. This is realised with a calculation of a loss function for the whole trajectory. It is based on a model of correct movement patterns, and a comparison of the recorded kinematics with the desired ones [7]. While designing a control system, adjusting its parameters, and training neural networks, the loss function may be also used to assess the performance of automatics. A schematic of the described control system is presented in Figure 2.

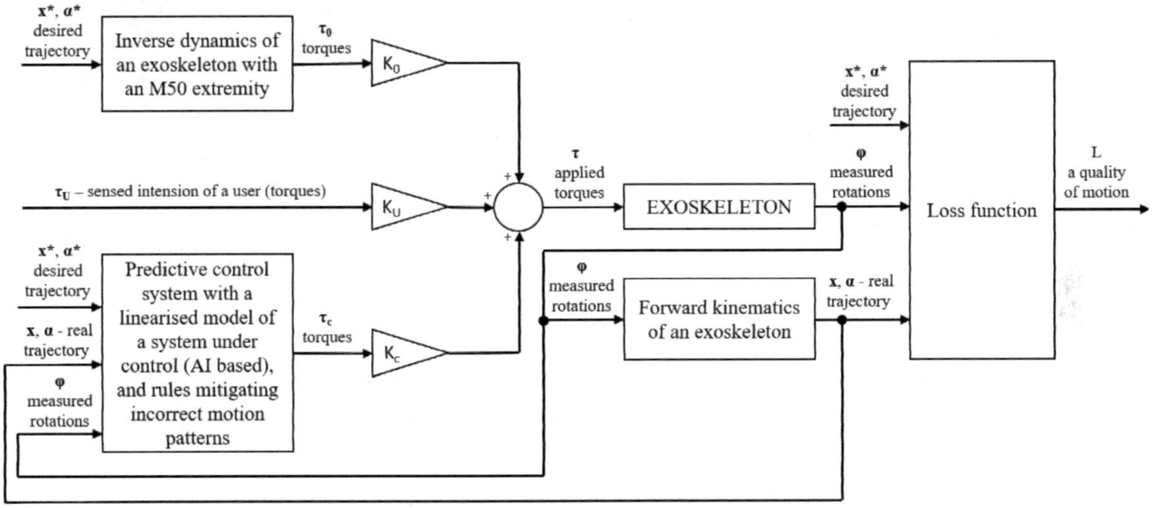

Figure 2. A schematic of a control system of the exoskeleton

The described system is aimed at performing operations maximally similar to the desired ones. As two DOFs remain free, a user is able to rotate them in an inaccurate way. To compensate for this, control algorithms have to adjust the work of three driven DOFs to perform tasks, follow the desired trajectory, and perceive anatomical kinematics patterns. Therefore, the loss function should consist of parts corresponding to the presented above.

3. Analysed motion patterns

Motion patterns of patients with selected diseases were analysed based on literature, consultations with physiotherapists and computational simulations. Afterwards, they were modelled mathematically and included in the loss function.

IOP Conf. Series: Materials Science and Engineering 1239 (2022) 012012 doi:10.1088/1757-899X/1239/1/012012

3.1. Trajectories for task-oriented rehabilitation.

The motion of an extremity is assessed only in terms of its kinematics. As for task-oriented training, reaching and manipulating movements were specified and analysed. These include drinking, eating, writing, steering a control wheel, lifting a bag from the ground, and lifting an object onto a shelf. Genuine motion patterns were compared to the ones common for post-stroke, neurologic, and orthopaedic patients. Based on this, typical motor compensations were defined and mathematically formulated.

3.2. Disability compensation.

While a neurologic or musculoskeletal system is affected, some motions may be impossible to perform. Patients typically try to compensate them with displacement and rotation of their trunk (as presented in Figure 3). A typical pattern of such is a usual shift of a shoulder girdle towards a reaching point [8]. Avoiding these is key in cognitive-motor training, as working out with incorrect technique may result in repeating incorrect learned movement even after the therapy. Eventually, may it lead to overstraining muscles and overloading joints, causing injuries and accelerated wear.

Figure 3. Incorrect motion patterns common for patients– bending forward in a sagittal plane to reach a point displaced far along a sagittal axis; rotating in a transverse plane to reach a point displaced far along a frontal axis, in the opposite direction to the reaching extremity; rotating in a frontal plane to reach a point displaced far along a longitudinal axis (from left)

3.3. Incorrect lifting pattern.

While lifting an upper extremity an arm should be aligned with a scapular plane [9]. However, motor disabilities may cause incorrect lifting patterns; so that, horizontal adduction is significantly bigger or smaller than required. Such a configuration of a limb does not exclude the possibility to lift it along the desired trajectory. Nevertheless, may it increase muscles tension, due to a non-anatomical motion pattern involved. If combined with spasticity, it may even end up with a break of muscle fibres.

4. Optimisation problem

An optimisation problem should enable an assessment of a real motion in regards to the desired pattern. As it is dedicated to task-oriented training, it needs to consider not only following a trail but also performing the desired operation and perceiving anatomical patterns, such as described in the previous section. To do so, distributions of components corresponding to different motion patterns and mathematical parameters have to be adjusted in order to raise a value of a function rapidly in case of emergency.

The symbols in the equations stand for: $\boldsymbol{x}$ – Cartesian trajectory vector; $\boldsymbol{\alpha}$ – angular trajectory vector; $\boldsymbol{\varphi}$ – inner variables (joint rotations) vector; t – time of an assessed movement; n – number of samples. Additionally, accents were added to mark: $\boldsymbol{v}^*$ - the desired $\boldsymbol{v}$ vector; $\hat{\boldsymbol{v}}_i$ – an element of a $\boldsymbol{v}$ vector associated with the closest point from the desired trajectory $\boldsymbol{x}_i^*$ regarding real current position $\boldsymbol{x}_i$; $\dot{\boldsymbol{v}}$ and $\ddot{\boldsymbol{v}}$ – the first and the second derivatives of a $\boldsymbol{v}$ vector.

4.1. Assessment of following a trajectory

Following a trajectory is the first aspect to be graded while assessing the motion. It should consist of components corresponding to differences between real and desired values of positions and angles, velocities and accelerations. Moreover, it should also consider the impact of time spent on realising an operation. A concept of a function to assess compliance with the desired motion pattern is presented in equation (1), where β are constants, which should be determined experimentally, once an exoskeleton is built.

$$T(\boldsymbol{x},\dot{\boldsymbol{x}},\ddot{\boldsymbol{x}},\boldsymbol{\alpha},\dot{\boldsymbol{\alpha}},\ddot{\boldsymbol{\alpha}},t)=\frac{1}{n}\sum_{i=1}^{n}\Big(\beta_1\|\boldsymbol{x}_i-\widehat{\boldsymbol{x}}_i^*\|^2+\beta_2\big\|\dot{\boldsymbol{x}}_i-\dot{\widehat{\boldsymbol{x}}}_i^*\big\|^2+\beta_3\big\|\ddot{\boldsymbol{x}}_i-\ddot{\widehat{\boldsymbol{x}}}_i^*\big\|^2+\beta_4\|\boldsymbol{\alpha}_i-\widehat{\boldsymbol{\alpha}}^*\|^2+ \beta_5\big\|\dot{\boldsymbol{\alpha}}_i-\dot{\widehat{\boldsymbol{\alpha}}}^*\big\|^2+\beta_6\big\|\ddot{\boldsymbol{\alpha}}_i-\ddot{\widehat{\boldsymbol{\alpha}}}^*\big\|^2\Big)+\beta_7\left(e^{\max\left(\frac{|t^*-t|}{t^*},\frac{t^*}{|t^*-t|}\right)}-1\right) \quad (1)$$

4.2. Penalties for incorrect motion patterns

Incorrect motion technique should affect a raise of the loss function correlated with the seriousness of a possible caused danger. While an angle between a scapular plane and the main axis of an arm φ_2 is relatively easy to calculate, compensation with trunk movements needs to be defined based on values of the registered inner variables. Therefore, rotating in a frontal plane may be detected by monitoring an angle between a longitudinal axis and the main axis of an arm ϕ (see formula (2)). Similarly bending forward in the sagittal plane may be detected by observing this, and an elbow's flexion angle φ_4. On the other hand, rotating in a transverse plane may be detected by observing an inner variable φ_3, corresponding to flexion in a shoulder joint. Moreover, penalty components corresponding to certain abnormalities in motion should be activated only under certain conditions, mainly dependent on a position of a characteristic point or its vertical velocity. Among others, the one corresponding to incorrect lifting pattern should be activated only for angular deviations from a model scapular plane larger than 5°, as the plane position may vary due to anatomical diversity. A concept of a penalty function for wrong motion patterns is presented in equation (3), where β are constants, which should be determined experimentally, once an exoskeleton is built, and l are values dependent on anatomical dimensions of a patient (see Figure 4).

$$\phi=\operatorname{asin}(\sin(\varphi_1)\cos(\varphi_2)) \quad (2)$$

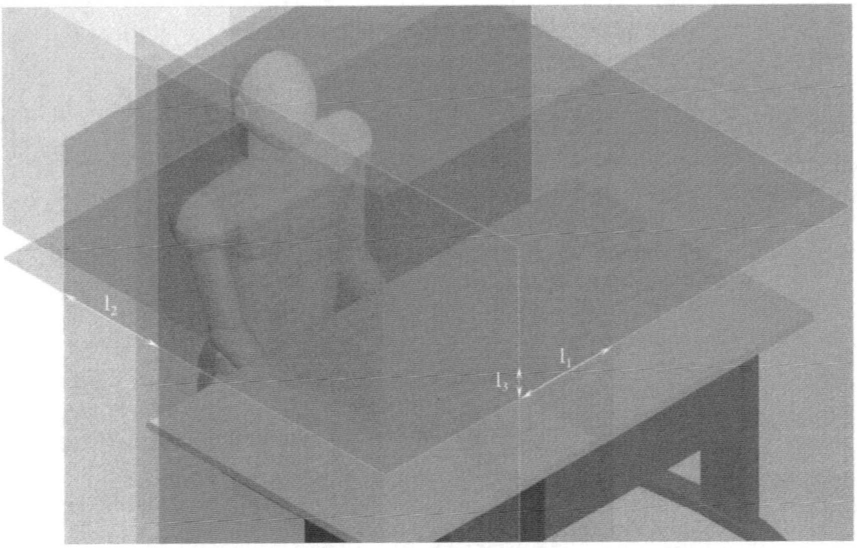

Figure 4. Activation regions (red for bending forward, yellow for rotating in a transverse plane, and green for rotating in a frontal plane) referring to the main planes (orange) of a shoulder joint in a neutral position

Components of the penalty formula (3) typically combine parts which activate them only if the characteristic point is in a certain activation zone (e.g. $\frac{-sgn(x_{1,i}+l_1)+1}{2}$) or a certain joint is out of range or below desired value (e.g. $\frac{sgn(\hat{\phi}_i^*-\phi_i)+1}{2}$), and parts responsible for a magnitude of penalty (e.g. $|\dot{x}_{3,i}|$ or $(\hat{\varphi}_{3,i}^* - \varphi_{3,i})^2$).

$$P(\boldsymbol{x},\boldsymbol{\varphi}) = \frac{1}{n}\sum_{i=1}^{n}\Bigg(\beta_9\frac{-sgn(x_{1,i}+l_1)+1}{2}\frac{sgn(\hat{\varphi}_{3,i}^*-\varphi_{3,i})+1}{2}\left(\hat{\varphi}_{3,i}^* - \varphi_{3,i}\right)^2 + \frac{sgn(\hat{\phi}_i^*-\phi_i)+1}{2}\left(\hat{\phi}_i^* - \phi_i\right)^2 \cdot \left(\beta_{10}\frac{sgn(x_{3,i}-l_3)+1}{2} + \beta_{11}\frac{sgn(x_{2,i}-l_2)+1}{2}\frac{sgn(\varphi_{4,i}-\hat{\varphi}_{4,i}^*)+1}{2}\left(\hat{\varphi}_{4,i}^* - \varphi_{4,i}\right)^2\right) + \beta_8|\dot{x}_{3,i}|\frac{sgn\left(\left(\varphi_{2,i}-\frac{5\pi}{36}\right)\left(\varphi_{2,i}-\frac{7\pi}{36}\right)\right)+1}{2}\left(\varphi_{2,i}-\frac{\pi}{6}\right)^2\Bigg) \quad (3)$$

4.3. Minimisation problem with constraints

A result of the two functions presented above is a loss function (4). It enables both, assessing a motion of a patient, and validating the quality of a created control system. An optimisation problem is a minimisation of the loss function, while the inner variables $\boldsymbol{\varphi}$ stay within anatomical ranges $\boldsymbol{\varphi}^{lim}$ presented in Table 1.

$$\min_{\boldsymbol{\varphi}\epsilon\boldsymbol{\varphi}^{lim}} L(\boldsymbol{x},\dot{\boldsymbol{x}},\ddot{\boldsymbol{x}},\boldsymbol{\alpha},\dot{\boldsymbol{\alpha}},\ddot{\boldsymbol{\alpha}},\boldsymbol{\varphi},t) = T(\boldsymbol{x},\dot{\boldsymbol{x}},\ddot{\boldsymbol{x}},\boldsymbol{\alpha},\dot{\boldsymbol{\alpha}},\ddot{\boldsymbol{\alpha}},t) + P(\boldsymbol{x},\boldsymbol{\varphi}) \quad (4)$$

5. Summary

The presented methodology enables automatic assessment of exoskeleton-aided rehabilitation. It is particularly important for the development of remote home-treatment devices. The proposed loss function not only compares the desired and the real trajectories but also analyses a technique of a patient's motion. Therefore, it is significantly useful in task-oriented workouts. Eventually, it may be applied for training artificial neural networks in predictive control algorithm, and as a part of a user grading system in a built exoskeleton.

6. References

[1] Y. K. Dwivedi, et al. Impact of COVID-19 pandemic on information management research and practice: Transforming education, work and life, International Journal of Information Management. 55 (2020).

[2] A. J. Bokolo, Use of Telemedicine and Virtual Care for Remote Treatment in Response to COVID-19 Pandemic, Journal of Medical Systems. 44, 132 (2020).

[3] F. J. Carod-Artal, Neurological complications of coronavirus and COVID-19, Revista de Neurologia. 70, 9 (2020) 311-322.

[4] D. C. Hess, W. Eldahshan, E. Rutkowski, COVID-19-Related Stroke, Translational Stroke Research. 11 (2020) 322–325.

[5] J. P. Bettger, L. J. Resnik, Telerehabilitation in the Age of COVID-19: An Opportunity for Learning Health System Research, Physical Therapy. 100, 11 (2020) 1913–1916.

[6] M. Marryam, M. Umar, Effectiveness of task oriented training in improving upper limb, Rawal Medical Journal. 42.3 (2017) 341-343.

[7] M. H. Lee, D. P. Siewiorek, A. Smailagic, A. Bernadino, S. B. Badia, Learning to assess the quality of stroke rehabilitation exercises, IUI'ig: Proceedings of the 24th International Conference on Intelligent User Interfaces. (2019) 218-228.

[8] S. Cai, et al. Detecting compensatory movements of stroke survivors using pressure distribution data and machine learning algorithms. Journal of NeuroEngineering and Rehabilitation 16, 131 (2019).
[9] N. A. Turpin, R. Martinez, M. Begon, Shoulder muscle activation strategies differ when lifting or lowering a load, European Journal of Applied Physiology. 120 (2020) 2417–2429.

IOP Conf. Series: Materials Science and Engineering 1239 (2022) 012013 doi:10.1088/1757-899X/1239/1/012013

Neural classifying system for predictive maintenance of rotating devices.

P Bałazy[1] , P Gut[1] and P Knap[1]

[1]AGH University of Science and Technology
Faculty of Mechanical Engineering and Robotics
Kraków, Poland

Abstract. This paper presents the results of research on a neural classifier system for the recognition of electrical drive disturbances and its speed. The classification task is performed by a convolutional neural network implemented on an artificial intelligence accelerator. The operation of the drive is monitored using intelligent accelerometers. Based on the sensor data, it is possible to create spectrograms of the signal using Short Time Fourier Transformation. Since convolutional neural networks are excellent at recognizing images and detecting objects, the signal spectrograms were used as input data for learning the network. Transfer learning method has been used, to create a model of neural network.

1. Introduction

Convolutional neural networks have proven to be an excellent solution for image recognition and object detection tasks over the past years [1]-[4]. Their growing popularity has led to newer and better network structures and newer areas of application. Most commonly, current draw, vibration, and temperature data are used in fault diagnosis of rotating equipment [5]-[9].

To perform vibration-based diagnosis, various signal analysis methods are used. The most common are fast Fourier transform (FFT), wavelet transform (CWT), and spectrogram representation of the signal by using (STFT). Especially the latter method is successfully used in combination with the use of convolutional artificial neural networks [10]-[13].

This paper discusses a proposal to use a convolutional neural network for device disturbance detection. The network was originally developed to classify images into a thousand categories. However, in the discussed application, it has been adapted using transfer learning for speed recognition and drive train disturbance detection.

2. Measurement station.

A test rig consisting of a servo drive and a geared motor was used. A single-axis piezoelectric accelerometer was used to measure acceleration, along with a conditioner and measurement card. A handle was designed and manufactured to simulate axial runout failure in the drive system, it is mounted on the shaft between the gearbox and the drive. The handle allows for the addition of additional mass that causes the system to be out of balance.

Content from this work may be used under the terms of the Creative Commons Attribution 3.0 licence. Any further distribution of this work must maintain attribution to the author(s) and the title of the work, journal citation and DOI.
Published under licence by IOP Publishing Ltd

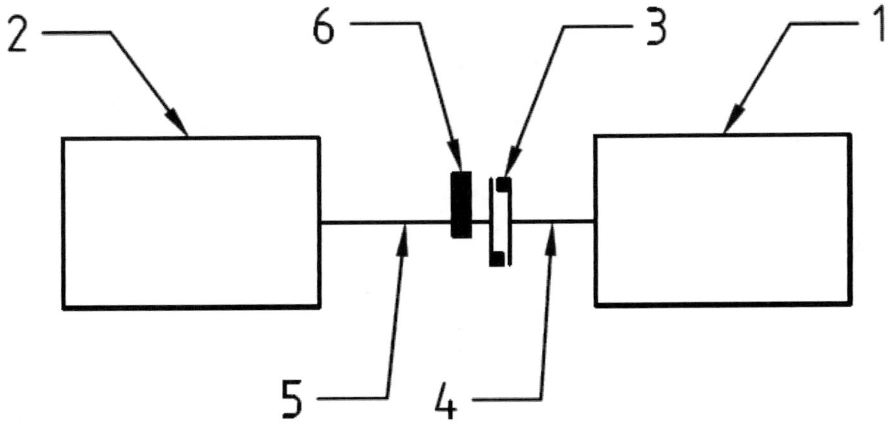

Figure 1. Drive system diagram.

In the schematic (figure 1) you can see the mounting location of the handle to simulate the anomalous operation of the system. The drive (1) drives the shaft (4), which is connected to the shaft (5) by a claw coupling with an elastomeric insert (3). The shaft (5) transmits the torque to the gearbox (2). During operation of the drive (1), the entire system emits vibrations, the accelerations of which are measured using an accelerometer on the gearbox housing (2). Abnormal operation is simulated by means of a handle (6) mounted on the shaft.

3. Data acquisition.

A single-axis ICP accelerometer was used to collect the necessary data along with a signal conditioner and an analog measurement card connected to a computer. The application developed using LabVIEW software allowed to save the read signals in TDMS type files. The view of the test stand is shown in (figure 2).

Figure 2. Stand for testing the condition of rotating machinery.

Four speeds 3000, 4000, 5000, and 6000 rpm were selected as the classes of the neural network. For each given speed, a twenty-minute vibration measurement of the unloaded system and with simulated abnormal operation was performed. The accelerometer was mounted in the mounting hole of the geared motor. The signal was sampled at a frequency of 1000 Hz.

IOP Conf. Series: Materials Science and Engineering 1239 (2022) 012013 doi:10.1088/1757-899X/1239/1/012013

4. Data pre-processing.

When a signal is analysed using the Fourier transform, the result is a frequency spectrum that only contains information about the relationship between the signal's frequency and its amplitude. The information about what happened to the signal over time disappears. The influence of time can be taken into consideration by using a spectrogram, where the horizontal axis represents time, the vertical axis represents frequency, and the intensity of coloring reflects the amplitude of a particular frequency in a given time interval. The spectrogram is a better solution also due to the nature of convolutional neural networks. It has more features that are easily recognizable to a neural network, such as colors, saturation, transitions between colors, and brightness. Below you can see a comparison of the frequency spectrum and the spectrogram for the same signal (figure 3) [4].

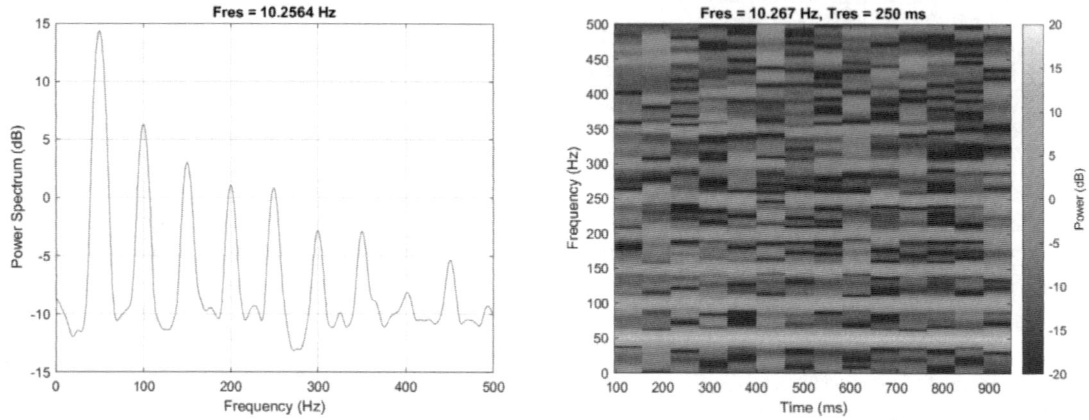

Figure 3. On the left frequency spectrum, on the right spectrogram of the measured signal.

A script was created in Matlab software to create images with spectrograms from the signal stored in TDMS files. Spectrograms are created from every thousand samples of the signal, which corresponds to one second of measurement. Spectrogram parameters, i.e. time resolution and color saturation scale, were selected experimentally. The script in each iteration creates a spectrogram and displays it, then the currently displayed graph is saved as a temporary file, which in the next iteration is overwritten. The saved image has axis descriptions and other information that is not needed. Therefore, it must be cropped so that only the spectrogram itself is visible. To do this, the saved image is loaded into the program, then cropped and saved as a PNG file. The image before cropping is at a resolution of 875x656, and after at 606x538 (figure 4). The cropping coordinates of the image were chosen experimentally. The result of the script is about 1200 images, numbered and ready to be used for training the neural network. For all categories a total of 10019 images were obtained.

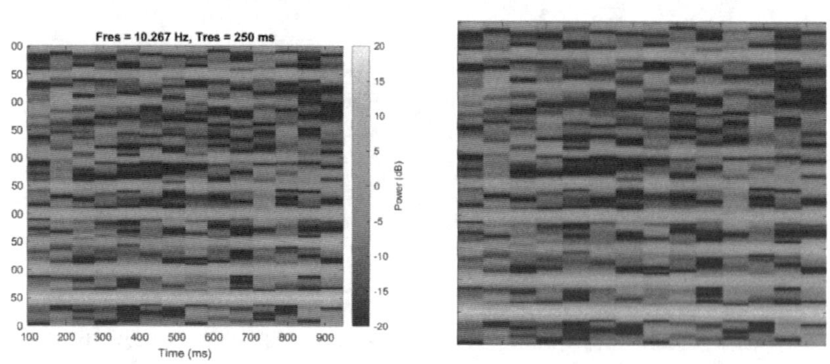

Figure 4. Sample spectrogram before and after crop.

5. Neural network training.

Transfer Learning was used to train the network, it is a commonly used approach in deep learning applications. It consists of choosing a ready, already trained model and using it as a starting point for solving your own problem. This approach also achieves satisfactory results using smaller datasets and shorter training time [14]-[16].

An already trained neural network called AlexNet was used. It is a convolutional neural network that has been trained on a collection of over a million images to classify thousands of object categories. AlexNet is characterized by simplicity and speed while maintaining satisfactory precision. It was necessary to adapt the structure of the selected network to the classification problem being solved. Changes were made in the last three layers of the network. The first of these layers is the fully connected layer, the number of neurons in it is defined by the number of classes. The next is the softmax layer, it is responsible for converting numerical values into probabilities of assignment to a given category. The last one is the classification layer, which is responsible for the assignment to one of the classes, and its output is the label of the class. The replacement of these layers is necessary because the network was trained to classify into a thousand different categories, and will now classify only eight. The input layer of the network is left unchanged. However, it requires 227x227x3 images, and the prepared spectrograms were 606x528x3, so it was necessary to scale them to the appropriate size.

The next step was to split the data set into learning, test, and validation sets. Many different divisions were tested, but the best results were obtained when 80% of all spectrograms were assigned to the learning set and 10% each to the test and validation sets. This corresponds to 8015 images in the learning set, 1003 images in the test set, and 1001 images in the validation set.

SGDM algorithm was chosen as the optimizer because it provided satisfactory precision and learning time. Due to hardware limitations, the batch size was assumed to be 64. The best model parameters were achieved for a learning rate of 0.0001. An important aspect is also the setting of the computing unit as a GPU, such a solution allows a significant reduction in the execution time of calculations

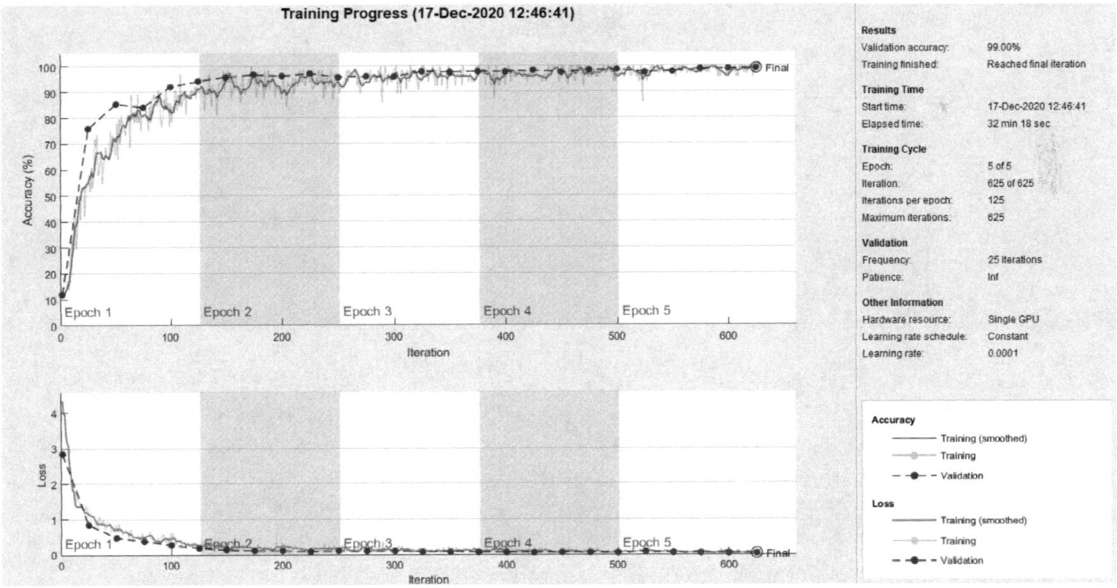

Figure 5. Progress achieved during training.

IOP Conf. Series: Materials Science and Engineering 1239 (2022) 012013 doi:10.1088/1757-899X/1239/1/012013

6. Results.

In the graph above (figure 5), you can observe the progress during learning. The neural network has already achieved quite good performance in the first epoch. In the fifth epoch, no more significant improvement in precision or minimization of the loss function can be seen. Finally, the following results were achieved:

- Precision on the learning set – 95.31%
- Value of the loss function for the learning set - 0.0937
- Precision on the validation set – 99,0%
- Value of the loss function for the validation set - 0.0352
- Precision on testing set – 99,1%

The above results are very satisfactory. The created confusion matrix (figure 6) for classification on the test set shows that the model was wrong only between two classes. It is also worth noting that the model achieves higher classification accuracy at some points on the validation set than on the test set. This is due to the use of a dropout layer in the AlexNet structure. During training, this layer randomly zeros the input values with a given probability, which in the case of the network used is 50%. This makes the learning of the network faster and protects it from overtraining. However, during testing or validation, the input output of the dropout layer is equal to its input.

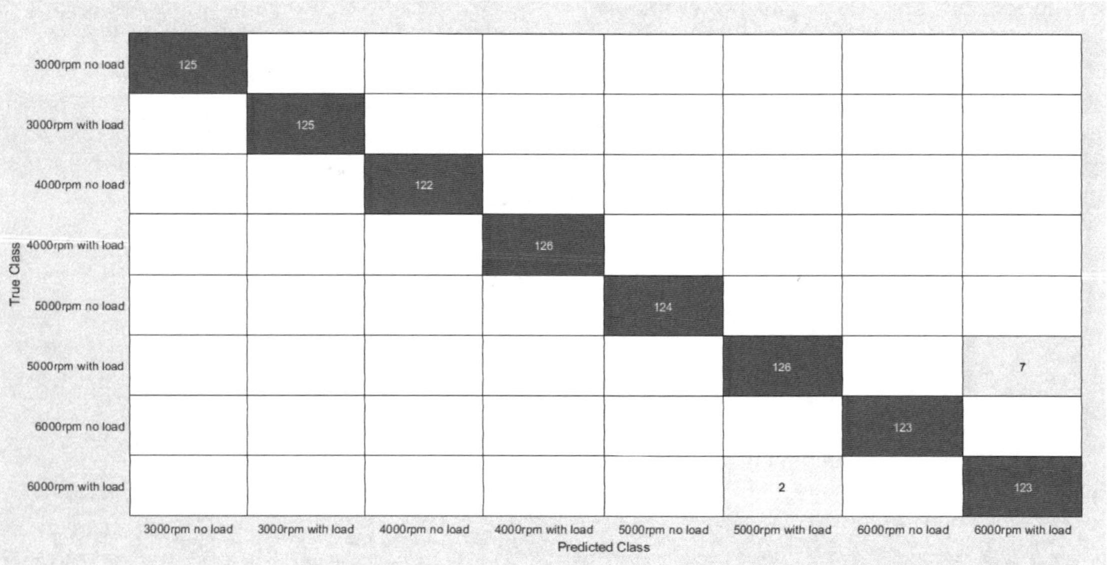

Figure 6. Confusion matrix chart.

7. Conclusion

The use of spectrograms as a database for the convolutional neural network has allowed for a very precise, efficient and relatively simple creation of a working classifier system for the analysis of the operation of rotating equipment. The proposed solution is characterized by a very high precision of detection of operating disturbances of the systems. This is confirmed by the fact that the designed classifier achieved a precision of 99.0% on the validation set and 99.1% on the testing set.

Designed solution perfectly fits into the strategy of Industry 4.0, which includes predictive maintenance. It assumes optimal use of machines, devices and other elements of infrastructure by eliminating failures and optimal planning of repair works or stoppages. After collecting a sufficient amount of data, it is possible to use the created algorithms to adapt them to work with other devices.

IOP Conf. Series: Materials Science and Engineering 1239 (2022) 012013 doi:10.1088/1757-899X/1239/1/012013

References

[1] A. Krizhevsky, I. Sutskever, and G. E. Hinton, "ImageNet classification with deep convolutional neural networks," *Commun. ACM*, 2017.

[2] D. A. Clevert, T. Unterthiner, and S. Hochreiter, "Fast and accurate deep network learning by exponential linear units (ELUs)," in *4th International Conference on Learning Representations, ICLR 2016 - Conference Track Proceedings*, 2016.

[3] K. Lalik and M. Kozek, "Cubic SVM neural classification algorithm for self-excited acoustical system," in *15th International Conference Mechatronic Systems and Materials, MSM 2020*, 2020.

[4] I. Dominik, J. Kwašniewski, and F. Kaszuba, "Ionic polymer-metal composite displacement sensors," *Sensors Actuators, A Phys.*, 2016.

[5] J. Zarei, M. A. Tajeddini, and H. R. Karimi, "Vibration analysis for bearing fault detection and classification using an intelligent filter," *Mechatronics*, 2014.

[6] I. Dominik, J. Kwasniewski, L. Krzysztof, and R. Dwornicka, "Preliminary signal filtering in Self-Excited Acoustical System for stress change measurement," in *Chinese Control Conference, CCC*, 2013.

[7] K. Lalik, I. Dominik, P. Ćwiąkała, and J. Kwaśniewski, "Integrated stress measurement system in tower crane mast," *Meas. J. Int. Meas. Confed.*, vol. 102, 2017.

[8] O. Janssens *et al.*, "Convolutional Neural Network Based Fault Detection for Rotating Machinery," *J. Sound Vib.*, 2016.

[9] I. Dominik, M. Iwaniec, and L. Lech, "Low Frequency Damage Analysis of Electric Pylon Model by Fuzzy Logic Application," *J. Low Freq. Noise Vib. Act. Control*, vol. 32, no. 3, pp. 239–251, 2013.

[10] Y. He, F. Luo, and M. Pan, "Defect characterisation based on pulsed eddy current imaging technique," *Sensors Actuators, A Phys.*, vol. 164, no. 1–2, pp. 1–7, 2010.

[11] W. Zhu, X. Li, C. Liu, F. Xue, and Y. Han, "An STFT-LSTM System for P-Wave Identification," *IEEE Geosci. Remote Sens. Lett.*, 2020.

[12] J. Huang, B. Chen, B. Yao, and W. He, "ECG Arrhythmia Classification Using STFT-Based Spectrogram and Convolutional Neural Network," *IEEE Access*, 2019.

[13] C. J. Ortiz-Echeverri, J. Rodríguez-Reséndiz, and M. Garduño-Aparicio, "An approach to STFT and CWT learning through music hands-on labs," *Comput. Appl. Eng. Educ.*, 2018.

[14] E. S. Olivas, J. D. M. Guerrero, M. M. Sober, J. R. M. Benedito, and A. J. S. López, "Handbook of Research on Machine Learning Applications and Trends: Algorithms, Methods, and Techniques," *Machine Learning in Natural Language Processing*. 2010.

[15] G. Zhao, Z. Zhang, J. Wang, and H. Guan, "Training Better CNNs Requires to Rethink ReLU," *arXiv*, 2017.

[16] M. Tan and Q. V. Le, "EfficientNet: Rethinking model scaling for convolutional neural networks," in *36th International Conference on Machine Learning, ICML 2019*, 2019.

IOP Conf. Series: Materials Science and Engineering 1239 (2022) 012014 doi:10.1088/1757-899X/1239/1/012014

Quality Research of Edge Banding of Unit Furniture

Ugnė Miškinytė [1], **Romualdas Juknelevičius**[1]

[1]Vilnius Gediminas Technical University (VILNIUS TECH), Saulėtekio al. 11, LT-10223 Vilnius, Lithuania

Abstract. Edge banding is used to cover cut sides of wood materials such as plywood or particleboard. The quality of two different furniture edge banding materials and their thickness were studied in this paper. The research was focused on most widely used in the Lithuanian furniture industry acrylonitrile butadiene styrene (ABS) and oak veneer. The experiments were performed to investigate the mechanical properties of materials and *SolidWorks Simulation* were made as well. The conclusions and recommendations can ensure the improvement and optimization of the processes of furniture industry. Structure of the paper: introduction, review of literature, methodology of experiment, results of research and their consideration, conclusions, references.

1. Introduction

The fast development of woodworking technology, was one of the reasons that unit furniture become more widely used in the furniture industry. According to the statistics of Lithuanian furniture industry of February – March the production of unit furniture increased by 24.9% [1] when compare 2020 and 2021. Edge lamination is used for the furniture made from wood imitation materials. The solid texture and colour of the furniture panel without any cracks should be ensured. The cracks are critical to furniture panels as the absorbed moisture can destroy the panel [2]. For this reason, the quality of laminating materials and the properties that determine it are very important for both manufacturers and their customers. Laminating tapes used for finishing the edges are: melamine, PVC, ABS, HPL, wood veneer. Although particleboard is an ideal basis for furniture, the manufacturer must pay attention to several key factors that will affect the quality [3].

Environmental protection is responsive issue of our Globe. The reduction of the number of materials used for production and use of recycled raw materials are effective solutions of sustainable production. One of the ways to reduce industrial waste is to use laminating tapes with respect to their quality.

In Lithuania, different thicknesses ABS and oak veneer edges are common in the furniture industry, therefore their quality is analysed. The study examines the strength, humidity, temperature and heat resistance of selected materials.

During the study of ABS and oak veneer edges were obtained results, which are suitable recommendations for the Lithuanian furniture industry and thus shorten the search for a suitable lamination tape, reduce the industrial waste and create a sustainable environment while saving natural resources.

The research was performed at the cabinet furniture company - *Evaldo baldai*. The following main problems of lamination process were identified at *Evaldo baldai*. These problems in fact are typical for the Lithuanian wood industry:

- Failure to evaluate the peeling force of the laminated edge.
- Lamination of curved parts.
- Resistance of strips to the factors of temperature differences.

Content from this work may be used under the terms of the Creative Commons Attribution 3.0 licence. Any further distribution of this work must maintain attribution to the author(s) and the title of the work, journal citation and DOI.
Published under licence by IOP Publishing Ltd

2. Review of literature

Humidity is one of the most important environmental factors influencing the quality of laminating tapes. However, humidity is directly related to temperature, another integral indicator of quality control. This statement is confirmed by a similar study analysing the quality of melamine bands in the Turkey. The study of Sacli [4] shows an increased bending force and tensile strength of the strip, and their change is related to changes in temperature.

Quality is also determined by the strength of the lamination tape, which can be checked when operating furniture at various temperatures. Such a study was conducted at the University of Warsaw [5], Institute of Wood and Furniture. It investigates the influence of temperature on selected strength properties of particle board.

A study by Naci and Nurgul Tankut [6] on the effects of compressive and tensile forces on laminating tapes describes the relationship between the forces acting on furniture and its quality. However, a clear relationship between edge band thickness and strength could not be established.

The summary of the reviewed scientific literature stated out that in order to better understand what parameters are important to compare in the mechanical and physical properties of laminating tapes, an analogous study applied to the materials used in Lithuania is needed. The matrix of the experiment, the research methodology, suitable for the edges of the ABS and oak veneer were determined after the literature overview.

3. Methodology of the research

The experiment and simulation was performed during the research. The *SolidWoks Simulation*, the mathematical FEM virtual modelling were applied. Analysed materials used in static; thermal analysis based on FEM. Particleboard 18 mm thick typical features are assigned to the model. The edges of the model are attributed to the properties of the lamination strips ABS and oak veneer. Dimensions of the model: width - 1000 mm, length - 600 mm. Simulations are performed by simulating real situations that occur during the operation of furniture.

The experiment of research was performed on EGGER particleboard. The board W908 SM was chosen and 16 pieces of samples were cut with machine *Format-4 kappa 400 x-motion* with dimensions of 100x60x18 mm. The cut parts were laminated with a BRANDT OPTIMAT KDF 230 C. EVA based *Kleiberit 788.7* adhesive is used for lamination. The cut specimens were divided into the groups according to the lamination tape with which they were laminated. A total of 4 different groups were obtained, laminated with: 22.00 mm wide and 2.00 mm thick ABS edge; ABS edge 22.00 mm wide and 0.80 mm thick; 24.00 mm wide and 2.00 mm thick oak veneer edge; 24.00 mm wide and 0.80 mm thick oak veneer edge. Before the tests, the samples were kept under laboratory conditions for 24 hours: at an average temperature of 18 °C and a relative humidity of 60 %. The following qualitative tests were performed: liquid resistance tests using absorbent medium, resistance to changes of an air humidity, resistance to temperature changes and peeling tests.

The obtained data were analysed, all research methods were summarized by providing conclusions and recommendations on the best use ABS and oak veneer lamination tapes in the furniture industry of Lithuania.

4. Results of research and their consideration

4.1. SolidWorks Simulation

4.1.1. Static analysis. The following analysis was performed using *SolidWoks Static Simulation*. The laminated part is reinforced through 3 edges. The shelf is fastened over 3 edges, simulates the position of the shelf in the wall niche and load of 25 kg was applied. The load F = 245 N. The results presented in the **Table 1**. According to the static analysis, in terms of quality, the edge of the oak veneer is stronger than ABS, and its strength increases with increasing strip thickness.

Table 1. Results of static analysis.

	ABS 0,8 mm	ABS 2 mm	Oak veneer 1 mm	Oak veneer 2 mm
Strength limit - maximum conditional stress (MPa)	0.551	0.499	2.061	1.916
Maximum displacement (mm)	0.731	0.733	0.714	0.702
Maximum deformation	$1.928 \cdot 10^{-4}$	$1.394 \cdot 10^{-4}$	$1.390 \cdot 10^{-4}$	$1.211 \cdot 10^{-4}$
Maximum stress N/mm² (MPa)	$5.516 \cdot 10^{-1}$	$4.990 \cdot 10^{-5}$	2.061	1.916
Minimum stress N/mm² (MPa)	$2.922 \cdot 10^{-5}$	$43.661 \cdot 10^{-5}$	$3.479 \cdot 10^{-5}$	$1.384 \cdot 10^{-4}$

4.1.2. Thermal analysis. Thermal analysis was performed with *SolidWoks Thermal Simulation*. The laminated part is exposed on all sides to the standard room temperature +23 °C, one side is exposed to the heat source of 600 W. Thus simulates the heat radiator of 1000 mm long and 400 mm high. The results of the virtual modeling of the thermal analysis (**Table 2**), stated out that the 2 mm thick edge of ABS is the most heat resistant. The least resistant is the 1 mm thick edge of the oak veneer. It is possible to distinguish the 1 mm laminated part of the oak veneer - retains the temperature the most, therefore this tape retains heat better than others.

Table 2. Results of thermal analysis.

	ABS 0,8 mm	ABS 2 mm	Oak veneer 1 mm	Oak veneer 2 mm
Maximum surface temperature (°C)	$7.825 \cdot 10^{4}$	$7.792 \cdot 10^{4}$	$7.829 \cdot 10^{4}$	$7.820 \cdot 10^{4}$

4.2. Experimental tests

4.2.1. Experiment of resistance to temperature changes. A practical experiment is presented to evaluate the resistance laminated samples to the temperature. It occurs during the transport of furniture in winter. Test conditions: the samples were kept in an oven at +60±3 °C for 8 hours. The samples were then immediately transferred to a freezer at –20±2 ° C for 16 hours. After the test, the samples were kept under standard ambient conditions for 24 hours. The results in **Figure 1** show that the absorbed heat flow of the oak veneer texture becomes brighter and the surface of the ABS edge fractures from the cold, but the deeper wood texture does not change the color of the edge, so in this respect the tree is superior to ABS with changing temperatures.

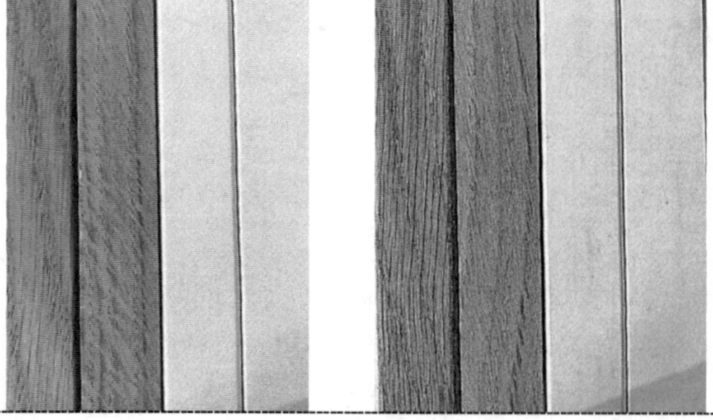

Figure 1. Temperature change experiment: samples on the left before the test, on the right after.

4.2.2. Thermal Analysis of resistance to liquids. The following Method was used with absorbent medium at elevated temperature. According to ISO 2812-3, varnishes, floor coverings and floor paints and interior finishes are water-resistant, if they do not lose their luster or discolor when exposed to water for 24 hours and 16 hours after the test. Test conditions: temperature +40±3 °C. The duration of the tests is 6 hours. The specimens were pressed with a moistened absorbent medium through the laminated edge. After the test, the samples were stored under standard ambient conditions for 24 hours. From the results in **Figure 2** ABS resistance to liquids is higher.

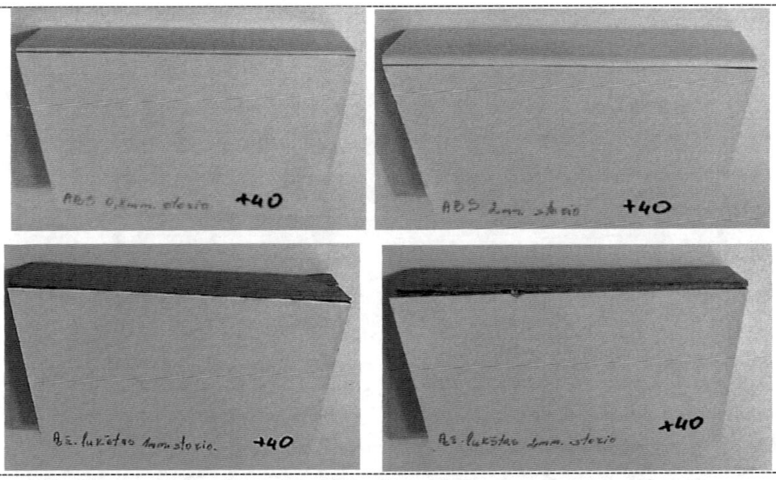

Figure 2. Temperature Results of the liquids resistance experiment:
1 - ABS 0.8 mm; 2 - ABS 2 mm; 3 - oak veneer 1 mm; 4 - oak veneer 2 mm.

4.2.3. Analysis of resistance to changes in humidity. Test conditions: the samples were conditioned for 8 hours at the temperature of +30±3 °C and a relative humidity of 95±5 %. The test conditions are max conditions of the bathroom. The samples were kept for 16 hours at the temperature of +23±2 °C and a relative humidity of 30±5 %. The visual changes are presented in Figure 3. The experiment proves that the resistance of ABS to humidity at elevated temperatures is higher, oak veneer does not tolerate high humidity, as there are clear changes in its appearance – any visual changes in the band after the experiment are considered quality discrepancies.

Figure 3. Results of the resistance to changes in humidity experiment:
1 - ABS 0.8 mm; 2 - ABS 2 mm; 3 - oak veneer 1 mm; 4 - oak veneer 2 mm.

IOP Conf. Series: Materials Science and Engineering 1239 (2022) 012014 doi:10.1088/1757-899X/1239/1/012014

4.2.4. Peel of test. 15 pcs. for each material were prepared. Test performed according to the standard LST EN 28510-1: 2014, peeling of glued flexible and rigid specimens together at an angle of 90 °. The peel forces are presented in Figure 4. From the test results presented in **Figure 4**, it can be stated that the results are distributed according to the normal distribution, which indicates the accuracy and reliability of the results. The results prove that the strongest material is oak veneer 2 mm. From the expanded samples (**Figure 5**) the strongest adhesion between the glue and the lamination edge is in 2 mm oak veneer material, and the weakest - 1 mm oak veneer.

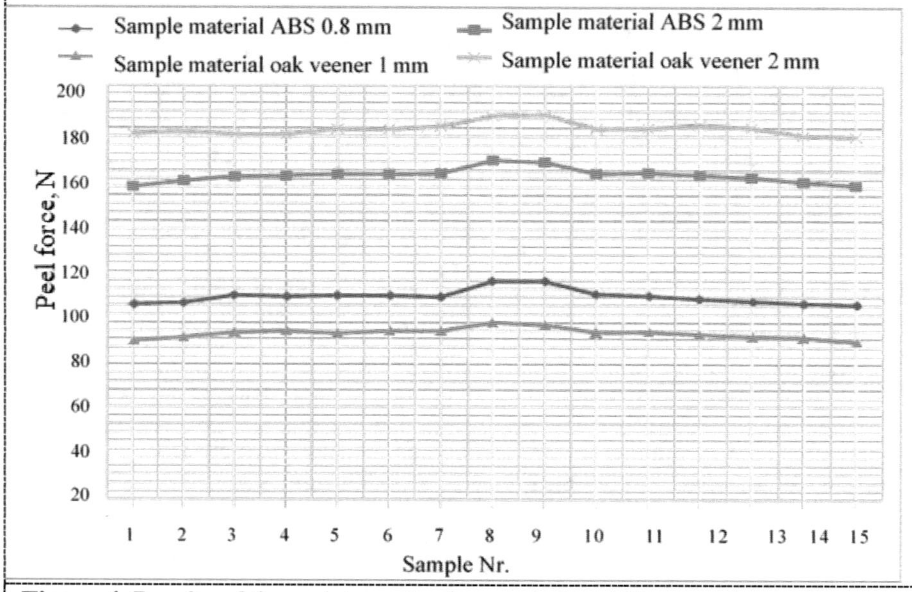

Figure 4. Results of the resistance to changes in humidity experiment:
1 - ABS 0.8 mm; 2 - ABS 2 mm; 3 - oak veneer 1 mm; 4 - oak veneer 2 mm.

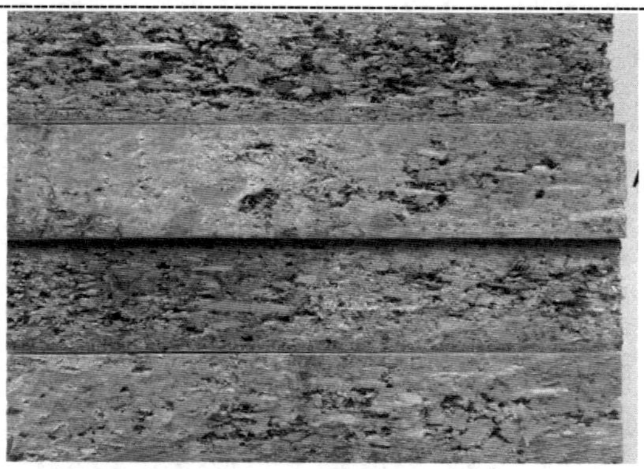

Figure 5. Result of particleboard after edges were peeled of (from the top):
1 - oak veneer 2 mm.; 2 - oak veneer 1 mm; 3 - ABS 2 mm; 4 - ABS 0.8 mm

IOP Conf. Series: Materials Science and Engineering 1239 (2022) 012014 doi:10.1088/1757-899X/1239/1/012014

5. Conclusions

After the experimental and simulation analysis of the lamination edges, the following conclusions and recommendations were conducted:

- An analysis of the literature has shown that the mechanical and physical properties of furniture, laminated with ABS and veneer edges, depend on the chemical composition of the materials and affect the durability of the product.
- According to the static analysis and experimental of the peeling force, the probability of deformation of furniture laminated with a thicker edge is decreasing. The adhesion of thicker edges to the particleboard is stronger.
- The lamination of curved shaped parts is affected by the elasticity of the edge. Since the modulus of elasticity of ABS is higher than of oak veneer, the application of plastic tape to such products would reduce the problem of internal radius lamination.
- Proposed recommendations for furniture operation for manufacturers:
  a. When choosing a laminating material, take into account their resistance to thermal factors that the furniture may face under foreseeable conditions in the future
  b. Evaluating the results of the peeling study, it can be considered that in furniture with a very frequent ratio of human strength to laminated edges, it is recommended to choose a thicker edge. It is suggested to choose oak veneer instead of ABS, due to stronger adhesion to the furniture panel.
  c. Based on the performed experiments of resistance to humidity and liquids, it is recommended to use ABS edge instead of oak veneer in furniture whose operation is associated with high humidity (outdoor and bathroom furniture).
  d. In terms of thermal permeability quality, the edge of the oak veneer is more suitable for use in furniture that provides close contact with a heat radiating device (furniture near a radiator, desk tops which computers).

6. References

[1] Lithuanian Department of Statistics 2020. *Changes in industrial production*

[2] Kajalavičius A 1981 *Baldžiaus vadovas* vol 2 (Vilnius: Mokslas)

[3] Nemli G Aydin I and Zekoviç E 2007 *Evaluation of some of the properties of particleboard as function of manufacturing parameters (Master Des.* p 1169–1176

[4] Sacli C 2015 *The effect of time and edge banding type and thickness on the bending and tensile strength of melamine coated particleboard* (*Research for Furniture Industry*) (Turkey/ Creative Commons Attribution) p 469–480

[5] Borysiuk P Tetelewska A Auriga R and Jenczyk I 2019 *The influence of temperature on selected strength properties of furniture particleboard (Annals of WULS Forestry and Wood Technology* vol 108) p 128–134

[6] Tankut, A and N 2010 *Evaluation the effects of edge banding type and thickness on the strength of corner joints in case–type furniture (Materials and Design* p 2956–2963). (Turkey/Bartin University, Faculty of Forestry, Department of Forest Industrial Engineering)

IOP Conf. Series: Materials Science and Engineering 1239 (2022) 012015 doi:10.1088/1757-899X/1239/1/012015

5DOF planar – rotary motion piezoelectric robot

A Čeponis[1], V Jūrėnas[1] and D Mažeika[2]

[1]Robotics and Piezomechanics Laboratory, Institute of Mechatronics, Kaunas University of Technology, K. Donelaičio str. 73, Kaunas, Lithuania
[2]Department of Information Systems, Faculty of Fundamental Sciences, Vilnius Gediminas Technical University, Saulėtekio al., 11, Vilnius, Lithuania

Abstract. A novel resonant type five degrees of freedom (5DOF) mobile piezoelectric robot is proposed and analysed. The piezoelectric robot performs unlimited linear working motion in the plane and rotates the positioning table or sphere about three axes independently. The piezorobot can be used for precise positioning or transportation of small objects and make complicated locomotion trajectories. The piezoelectric robot is designed as a compound structure and includes a piezoelectric ring and a special ring-shaped bronze layer. Electrodes of the piezoelectric ring are divided into six equal sections. Excitation of the piezoelectric robot is performed by using a single switched harmonic signal applied to the particular electrode. The operation principle of the piezoelectric robot is based on the excitation of the third radial vibration mode of the piezoelectric ring. The numerical and experimental study was performed, and the operating principle was validated. The maximum linear velocity of 18.8 mm/s and rotational speed of 31.3 rpm were achieved when load of 25.1g and excitation voltage of 200 V_{p-p} were applied.

1. Introduction

Modern high-precision mechatronic systems include different types of actuators and motors, such as piezoelectric, magnetostrictive, electrostatic, electromagnetic, etc. Piezoelectric devices have advanced features and can be used as sensors and actuators at the same time because of the direct and indirect piezoelectric effect [1]. Mechanical vibrations of the actuator excited by electric signal can be transferred into continuous linear or rotational motion of the slider or rotor [2]. Also, the actuator can move itself on the plane. These unique features allow the development of multipurpose mobile piezoelectric robots that can manipulate the positioning object, move into a particular location with high accuracy and perform different tasks. Piezoelectric robots have a simple design, good controllability, and can achieve high resolution [3]. The size of piezo robots can be minimized up to a millimetric scale. Vibration amplitude and waveform of mechanical vibration of the contacting legs can be easily controlled by electric signals. The electric circuit can be configured to actuate or damp mechanical vibrations, therefore a wide range of the control parameters are used to control the piezoelectric robot in a proper way.

Moreover, multi-degrees of freedom (MDOF) piezoelectric robots can be designed using a single piezoelectric actuator. The structural design of the robot and electrode topology must be optimized to construct MDOF piezoelectric robot. The piezoelectric actuator must generate the elliptical oscillation trajectories of the contact points or impact contact surface at an inclined angle in order to obtain the locomotion of the robot. The inertial driving principle can be used to drive robot by employing

Content from this work may be used under the terms of the Creative Commons Attribution 3.0 licence. Any further distribution of this work must maintain attribution to the author(s) and the title of the work, journal citation and DOI.
Published under licence by IOP Publishing Ltd

asymmetric vibrations. Elliptical motion can be generated using mode superposition, shifted phase excitation, or vibro-impact principle, while asymmetric vibrations of the contact point is obtained applying square or triangle waveform electric signal. MDOF ultrasonic piezoelectric systems face the problem of the excitation of the particular single degree of freedom independently. The whole solid system vibrates when any zone of the actuator is excited. It complicates control of MDOF piezoelectric robot locomotion, and a closed-loop feedback system, shunting, and out-of-phase excitation are required.

There are reports on piezoelectric robots used for planar motion and carry a payload [4-8]. Robots have different designs and are based on the cylinder, disc, or plate type bimorph piezoelectric transducers. It was shown that disc and cylinder type piezoelectric robots can achieve controllable 2DOF planar motion and rotate about the vertical axis [4, 5]. Robots with plate-type bimorph actuators can provide linear locomotion [6, 7] or planar motion [8]. However, there is no piezoelectric robot with a single actuator that could provide planar locomotion as well as rotate payload about two axes.

This study presents a novel design of a 5DOF piezoelectric robot. Numerical analysis based on finite element modeling was performed, and the operating principle was validated. A prototype motor was made, and an experimental study confirmed the feasibility of the robot. Finally, the corresponding conclusions are made.

2. Design and operation principle

The piezoelectric robot is designed as a compound structure and consists of a piezoelectric ring and ring-shaped metal layer (Fig.1). Both parts are glued using epoxy. The top electrode of the piezoelectric ring is divided into six equal sections, while the bottom electrode is common and used for grounding. The polarization direction of the ring is aligned with the thickness of the ring. The ring-shaped metal layer is divided into the same six sections as the piezoelectric ring. The thickness of the ring at the two adjacent sections is different. This was done to make the vibration amplitudes of the adjacent sections different when one of the sections is excited by an electric signal. Moreover, each section of the metal ring is separated by gaps to minimize vibration energy transfer to the neighboring sections. Totally six alumina spherical balls are used as contacting structures between the robot and surface (Fig 1). Three balls are glued on the top surface of the piezoelectric ring and are used for rotation of a sphere, while the other three balls are glued at the bottom and are employed to obtain the planar motion of the robot. All balls were placed in the middle of the corresponding section.

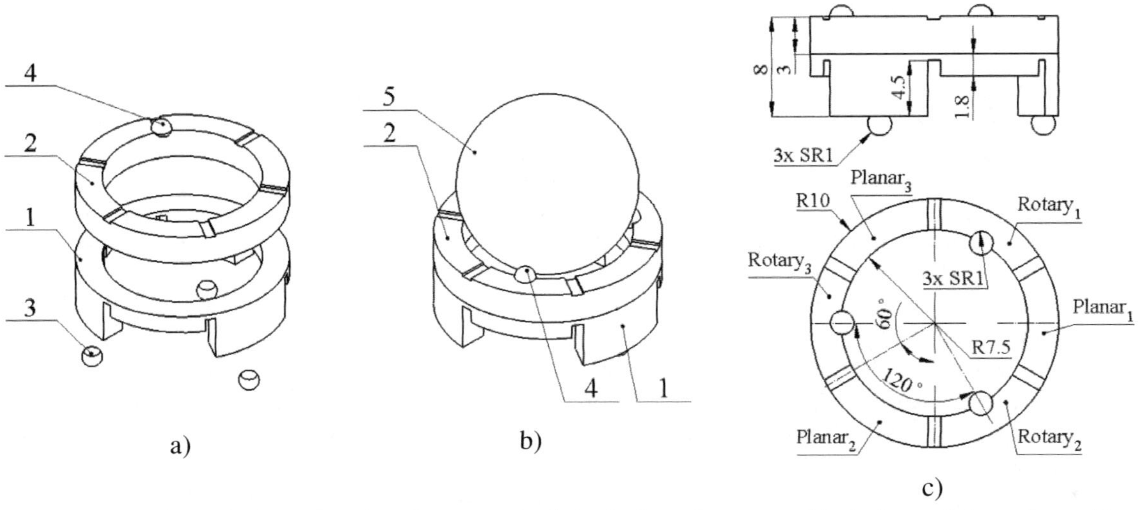

Figure 1. Design of piezoelectric robot: a - exploited view; b - assembled view of the robot with a sphere; c – sketch of the robot. 1 – bronze ring; 2 – piezoceramic ring; 3 – spherical contacts for planar motion; 4 – spherical contacts for rotary motion; 5 – spherical rotor.

The operation of the robot is based on the 3rd radial vibration mode of the ring-shaped structure using d_{31} piezoelectric effect. In order to implement flexible control of a particular degree of freedom, the structure of the robot is divided into six sections. Excitation of the piezo robot is performed by using a single switched harmonic signal applied to the particular electrode. Such individual excitation of the electrode allows achieving planar or rotary motion in an appropriate direction.

Locomotion control of the robot as well as control of the sphere rotation, can be implemented using digitally controlled switching and a single harmonic signal. Figure 2 shows excitation schematics of the robot using a single generator and two switching boxes used to control payload rotary motion and planar motion of the robot, respectively. Locomotion along the X axis is obtained by switching on SW5, and reverse motion is generated by switching SW4 and SW6. Movement in the Y direction is generated by switching SW5 and SW6, while reverse motion is obtain using SW4 and SW5. A similar switching principle is applied to obtain rotation of the sphere about the X and Y-axis. In order to achieve rotation about the Z-axis, three signals shifted by 120 degrees must be used. Special trajectory planning algorithms can be used to generate the required planar or rotational motion [4]. The amplitude of the electric signal, duration time, and sequences of switching control allows achieving needed motion of the piezoelectric robot and payload. High displacement and rotation resolution of the robot can be achieved by applying burst type electric signal.

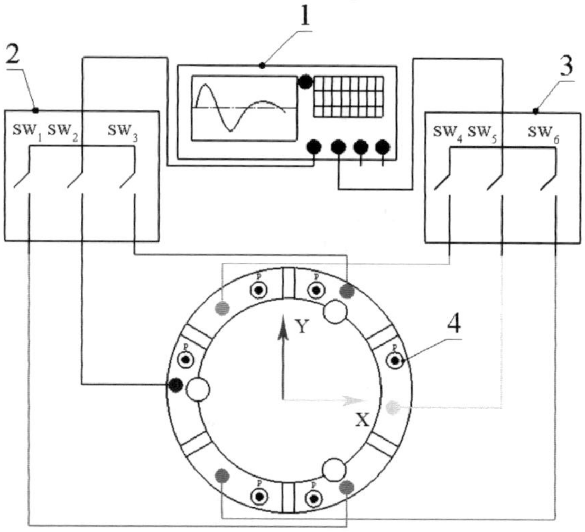

Figure 2. Excitation schematics of the robot: 1 – generator, 2 – switching box for control of payload rotary motion 3 – switching box for control of planar motion, 4 – electrode and polarization direction.

3. Numerical modelling

Numerical calculations of the piezoelectric robot was performed to analyse modal shapes and make frequency response analyses. The finite element model of the robot was built using Comsol Mutiphysics 5.4. The following materials were used i.e., beryllium bronze was used for the passive layer of the robot, PZT8 piezo ceramic was used for the piezoelectric ring, and alumina oxide was used for contacting balls. The adhesive layer was neglected in the model. The model of the robot was not clamped while all electrodes were grounded when modal frequency analysis was performed.

3.1 Results

The operational modal shape of the piezoelectric robot is shown in Figure 3a when it vibrates at the frequency of 58.085 kHz. The 3rd radial vibration mode is dominating in this shape. The anti-nodes of the modal shape are located in the sections used for planar motion, while sections used for rotation are in the nodes of modal shape. The impedance of the robot was analysed in a frequency range of 57.60 – 58.20 kHz in order to analyse the losses of the system (Fig. 3b). It can be seen that the lowest impedance value of 1.19 kΩ is obtained at the resonant frequency of 58.01 kHz. Based on the impedance graph mechanical quality factor Q_m of the robot was calculated and it was found that Q_m = 1055.54. The effective coupling coefficient k_{eff} is 0.0435. These results show that the piezoelectric robot has low mechanical loss at the operating frequency.

Harmonic response analysis was performed by analysing displacement amplitudes of the six spherical contacts in the frequency range from 58.0 kHz till 58.2 kHz with the step of 2.5 Hz when the voltage of 200 V_{p-p} was applied to a single electrode. This type of excitation evaluates planar and rotational possibilities in one direction (Fig. 4). Calculations were performed when the robot was preloaded by the force of 0.546N. Results of simulation show that a peak of displacement amplitude of 3.88 μm is obtained at the frequency of 58.088 kHz. This peak value corresponds displacement amplitude of the spherical contacting ball located in the section that was excited by an electric signal. It must be noted that other contacting balls have significantly lower amplitudes of vibrations.

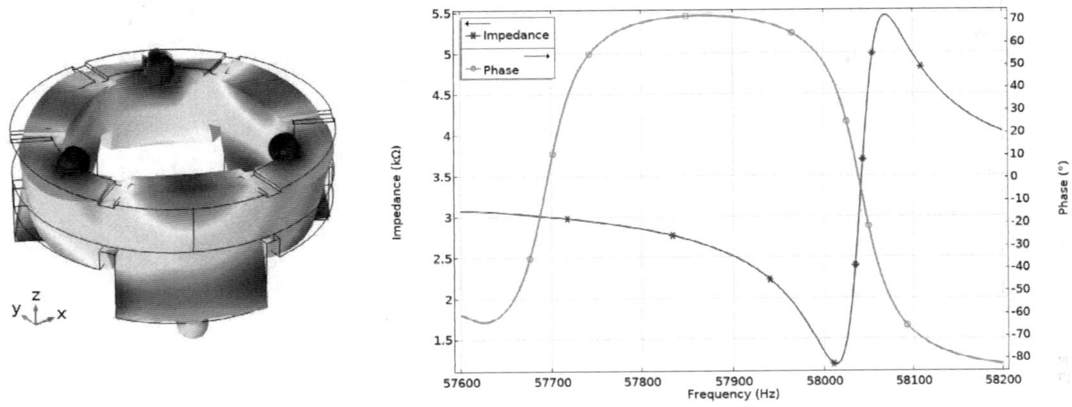

a) b)

Figure 3. The modal shape of the actuator at 58.085 kHz (a) and dependence of impedance and phase versus frequency (b).

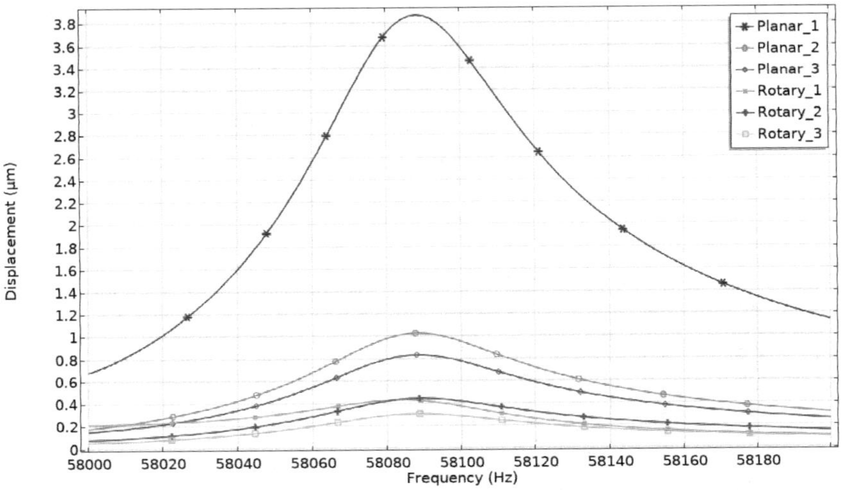

Figure 4. Amplitude – frequency characteristics of the different contact points of the robot.

It was already mentioned in the introduction that the MDOF ultrasonic piezoelectric systems face the problem of the excitation of the particular single degree of freedom independently because the whole system vibrates when any zone of the actuator is excited. Therefore, a more detailed numerical analysis was performed to compare the amplitudes of the vibrations of the spherical contacts located in the excited zone and the amplitudes of the remaining contacting balls. The particular single electrode was excited by the harmonic voltage of 200 V_{p-p}, and amplitudes of all spherical contacts were analysed. The results are shown in figure 5. It can be seen that difference between vibrations amplitudes of the spherical contacts used for planar motion is up to 4.02 times and planar – rotational is up to 12.9 times (Fig.5a). A similar comparison was made between the vibration amplitudes of the spherical contacts used for rotary motion and the remaining spherical contacts. It was found that the difference is up to 4.3 times and 11.2 times, respectively. Obtained results show that the influence of the contacting spheres located on non-excited sections makes a small influence on the driving direction defined by the exciting section. Therefore, it can be concluded that the direction of locomotion or rotation of the spherical payload can be controlled by applying an electric signal to the corresponding electrode.

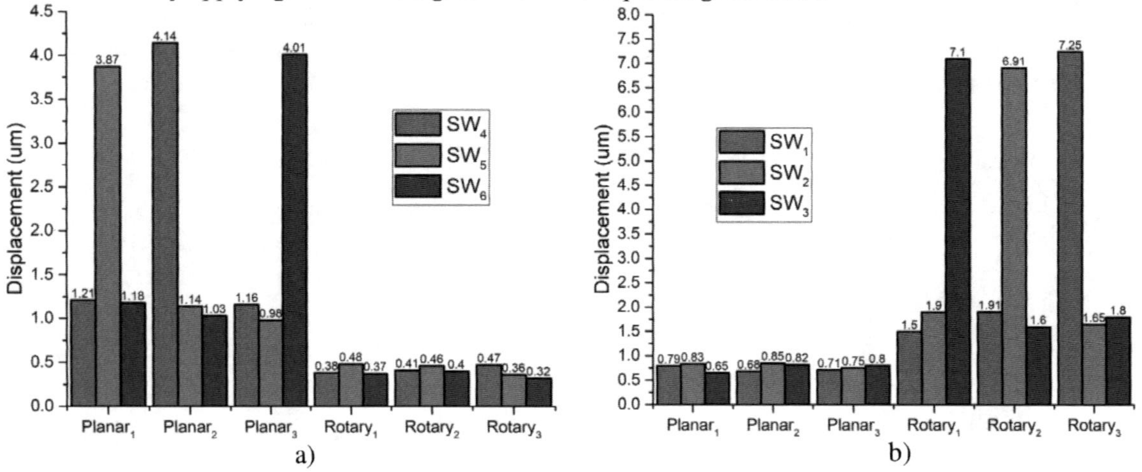

Figure 5. Comparison of the vibration amplitude of the spherical contacts used for planar motion (a) and rotational motion (b).

4. Experiments

An experimental study of the piezoelectric robot was performed to validate the operating principle and to measure electrical and mechanical characteristics. A prototype was made and is shown in figure 6a. Impedance–frequency characteristic of the robot was measured without preload. Analysing results, it can be seen that resonant frequency was obtained at the frequency of 59.65 kHz (Fig. 6b). The difference between calculated and measured frequency is 2.68%. The measured mechanical quality factor Q_m is 950.12 while the effective coupling coefficient k_{eff} is 0.058. The difference does not exceed 2.5% compared with the results obtained from the numerical study.

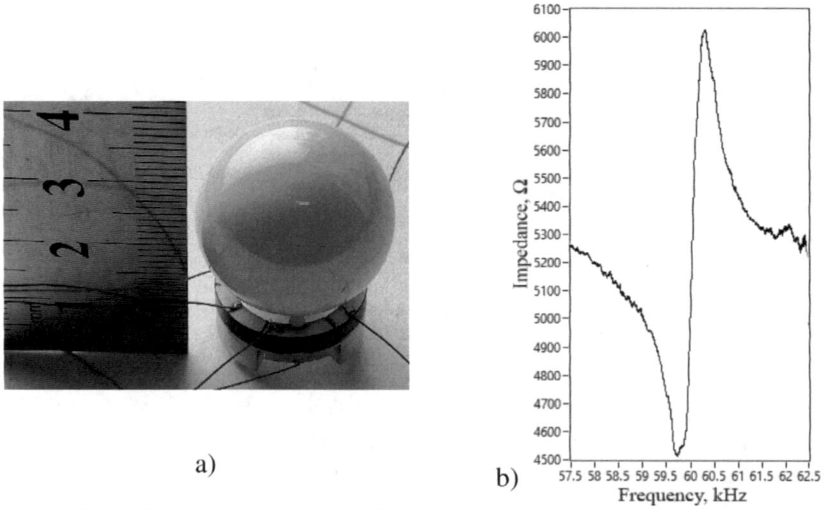

Figure 6. Prototype of the piezoelectric robot with a sphere (a), the measured impedance of piezoelectric robot in the frequency domain (b).

Measurements of the linear and rotation velocity were performed when different preload and voltage was applied. The preload force was obtained using spheres with different weights. Figure 7 shows average linear and rotation velocity when the voltage from 80 V_{p-p} to 200 V_{p-p} was applied. It can be seen that velocity almost linearly depends on the driving voltage and is increasing with the increase of preload. This can be explained by the fact that as the load increases, the friction force between contacting balls and planar surfaces increases, so the robot moves at a higher speed. The maximum average linear speed of 18.8 mm/s was obtained at the preload of 0.251 N while the voltage of 200V_{p-p} was applied. Accordingly, the highest rotation speed of 31.3 RPM was obtained at the same conditions.

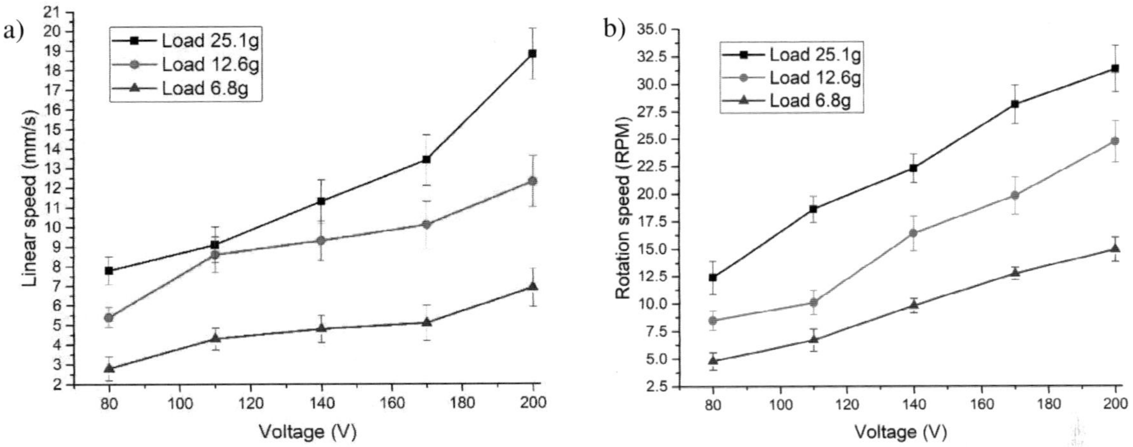

Figure 7. Measured average linear (a) and rotation (b) velocity of the piezoelectric robot

5. Conclusions

A novel 5-DOF piezoelectric robot was developed and investigated. Results of the numerical and experimental study confirmed that the robot can perform planar locomotion and rotary motion of the payload. Motion control is performed by switching the harmonic signal applied to the six electrodes of the piezoelectric ring. The maximum average rotation speed of 31.3 RPM and linear velocity of 18.8 mm/s was obtained during the experimental study when the voltage of 200 V_{p-p} was applied.

Acknowledgment

This research was funded by the European Social Fund according to the activity "Promotion of postdoctoral fellowships studies" of Measure No. 09.3.3-LMT-K-712-19-0082.

References

[1] Uchino K, Giniewicz J 2003 *Micromechatronics* New York Marcel Dekker Inc.

[2] Zhao C 2011 *Ultrasonic motors: Technologies and applications* Beijing Springer Science Press.

[3] Behera B, Nemade H B 2017 Recent developments of piezoelectric motors with diverse operating principles. *Journal of Micro and Smart Systems* vol. 6(2) p 173-185.

[4] Janutėnaitė-Bogdanienė J, Macerauskas E, Drukteiniene A, Kulvietis G, Bansevicius R. Cylindrical piezorobot's trajectory planning and control 2017 *J. of Vibr.* **19(4)**, p 2670-2679.

[5] Bansevičius R, Drukteiniene A, Kulvietis G 2010 Movement Trajectory Planning Algorithm of Rotating Mobile Piezorobot. *Solid State Phenomena*, **164**, p 371–376.

[6] Hariri H H, Soh G S, Foong S, Wood K L 2019 A Highly Manoeuvrable and Untethered Under-Actuated Legged Piezoelectric Miniature Robot. *Proc. ASME 2019 Int. Design Engineering Technical Conferences and Computers and Information in Engineering Conference* Anaheim, California, USA. August 18–21 2019 vol. 5B: V05BT07A004.

[7] Hernando-Garcia J, Garcia-Caraballo J L, Ruiz-Diez V, Sánchez-Rojas J L 2020 Motion of a Legged Bidirectional Miniature Piezoelectric Robot Based on Traveling Wave Generation. *Micromachines* **11(3)** 321.

[8] Hariri H, Bernard Y, Razek A 2018 2-D Traveling wave driven piezoelectric plate robot for planar motion. *IEEE/ASME Transactions on Mechatronics* **23** p 242–251.

AUTHOR INDEX